COTTON IS KING:

OR THE

CULTURE OF COTTON, AND ITS RELATION TO

Agriculture, Manufactures and Commerce;

TO THE

FREE COLORED PEOPLE; AND TO THOSE WHO HOLD
THAT SLAVERY IS IN ITSELF SINFUL

BY AN AMERICAN.

David Christy

CINCINNATI:
MOORE, WILSTACH, KEYS & CO.,
25 WEST FOURTH STREET.
1855.

WM. OVEREND & CO., Printers,
CINCINNATI.

PREFACE.

In the preparation of the following pages, the Author has aimed at clearness of statement, rather than elegance of diction. He sets up no claim to literary distinction; and, even if he did, every man of classical taste knows, that a work, abounding in facts and statistics, affords little opportunity for any display of literary ability.

The greatest care has been taken, by the Author, to secure perfect accuracy in the statistical information supplied and in all the facts stated.

The authorities consulted, are Brande's Dictionary of Science, Literature and Art; Porter's Progress of the British Nation; McCullough's Commercial Dictionary; Encyclopædia Americana; London Economist; De Bow's Review; Patent Office Reports; Congressional Reports on Commerce and Navigation; Abstract of the Census Reports, 1850; and Compendium of the Census Reports. The extracts from the

Debates in Congress, on the Tariff Question, are copied from the *National Intelligencer*.

The tabular statements appended, bring together the principal facts, belonging to the questions examined, in such manner, that their relations to each other can be seen at a glance.

The first of these Tables, shows the date of the origin of Cotton Manufactories in England, and the amount of Cotton annually consumed, down to 1853; the origin and amount of the exports of Cotton from the United States to Europe; the sources of England's supplies of Cotton, from countries other than the United States; the dates of the discoveries which have promoted the production and manufacture of Cotton; the commencement of the movements made to meliorate the condition of the African race; and the occurrence of events that have increased the value of slavery and led to its extension.

The second and third of the tables, relate to the exports and imports of the United States; and illustrate the relations sustained by slavery, to the other industrial interests and the commerce of the country.

CONTENTS.

II.

III.

APPENDIX.

COTTON IS KING.

THE controversy on SLAVERY in the United States, has been one of an exciting and complicated character. The power to *emancipate* existing, in fact, in the States separately, and not in the General Government, the efforts to abolish it, by appeals to public opinion, have been fruitless, except when confined to single States. In Great Britain, the question was simple. The power to abolish Slavery in her West Indian colonies was vested in Parliament. To agitate the people of England, and call out a full expression of sentiment, was to control Parliament, and secure its abolition. The success of the English Abolitionists, in the employment of moral force, had a powerful influence in modifying the policy of American Anti-slavery men. Failing to discern the difference in the condition of the two countries, they attempted to create a public sentiment throughout the United States, adverse to Slavery, in the confident expectation of speedily overthrowing the institution. The issue taken, that Slavery is *malum in se*—a sin in itself—was prosecuted with all the zeal and eloquence they could command. Churches, adopting the *per se* doctrine, inquired of their converts, not whether they supported Slavery, by the use of its products, but whether they believed the institution itself sinful. Could public sentiment be brought to assume the proper ground; could the Slaveholder be convinced that the world denounced him as equally criminal with the robber and murderer; then, it was believed, he would abandon the system. Political parties, subsequently organized, taught, that to vote for a Slaveholder, or a Pro-slavery man, was sinful, and could not be done without violence to conscience; while, at the same time, they made no scruples of using the products of Slave labor—

the exorbitant demand for which was the great bulwark of the institution. This was a radical error. It laid all who adopted it open to the charge of practical inconsistency, and left them without any moral power over the consciences of others. As long as all used their products, so long the Slaveholders found the *per se* doctrine working them no harm; as long as no provision was made for supplying the demand for tropical products, by free labor, so long there was no risk in extending the field of their operations. Thus, the very things necessary to the overthrow of American Slavery, were left undone, while those essential to its prosperity, were continued in the most active operation; so that, now, after nearly a "thirty years' war," we may say, emphatically, COTTON IS KING, and his enemies are vanquished.

Under these circumstances, it is due to the age—to the friends of humanity—to the cause of liberty—to the "safety of the Union"—that we should review the movements made in behalf of the African race, in our country; so that errors of principle may be abandoned;

mistakes in policy corrected; incompetent leaders discharged; the free colored people induced to change their relations to the industrial interests of the world; the rights of the slave, as well as the master, secured; and the principles of our Constitution established and revered. We propose, therefore, to examine this subject, as it stands connected with the history of our country; and especially to afford some light to the free colored man, on the true relations he sustains to African Slavery, and to the redemption of his race. The facts and arguments we propose to offer, will be embraced under the following heads:

I. The circumstances under which the American Colonization Society took its rise; the relations it sustained to Slavery, and to the schemes projected for its abolition; the origin of the elements which have given to American Slavery its commercial value, and consequent powers of expansion; and the futility of the means used to prevent the extension of the institution.

II. The present relations of American Slavery to the Industrial interests of our own country; to the demands of Commerce; and to the present Political crisis.

III. The social and moral condition of the free colored people in the British Colonies, and in the United States; and the new field opening in Liberia, for the display of their powers.

IV. The moral relations of persons holding the *per se* doctrine, on the subject of Slavery, to the purchase and consumption of Slave labor products.

1. Four years after the Declaration of American Independence, Pennsylvania and Massachusetts had emancipated their slaves; and, eight years thereafter, Connecticut and Rhode Island followed their example.

Three years after the last-named event, an *Abolition Society* was organized by the citizens of the State of New York, with JOHN JAY at its head. Two years subsequently, the Pennsylvanians did the same thing, electing BENJAMIN FRANKLIN to the presidency of their association. The same year, too, Slavery was forever excluded, by Act of Congress, from the North-west territory.

During the year that the New York Abolition Society was formed, WATTS, of England, had so far perfected the *steam engine* as to use it in propelling machinery for spinning cotton; and the year the Pennsylvania Society was organized, witnessed the invention of the *Power Loom.* The *Carding Machine* and the *Spinning Jenny* having been invented twenty years before, the Power Loom completed the machinery necessary to the indefinite extension of the manufacture of cotton.

The work of emancipation, begun by the four States named, continued to progress, so that, in seventeen years from the adoption of the Constitution, New Hampshire, Vermont, New York, and New Jersey, had also enacted laws, to free themselves from the burden of Slavery.

As the work of manumission proceeded, the elements of Slavery expansion were multiplied. When the four States first named liberated their slaves, no regular exports of cotton to Europe had yet commenced; and the year New Hampshire set hers free, only 138,328 pounds of that article were shipped from the country. Simultaneously with the action of Vermont, in the year following, the *Cotton Gin* was invented, and an unparalleled impulse given to the cultivation of cotton. At the same time, Louisiana, with her immense territory, was added to the Union, and room for the extension of Slavery vastly increased. New York lagged behind Vermont for six years, before taking her first step to free her slaves, when she found the exports of cotton to England had reached 9,500,000 pounds; and New Jersey, still more tardy, fell five years behind New York; at which time the exports of that staple — so rapidly had its cultivation progressed, were augmented to 38,900,000 pounds.

Four years after the emancipations, by States, had ceased, the Slave trade was prohibited; but, as if each movement for freedom must have its counter-movement to stimulate Slavery, that same year the manufacture of cotton goods was commenced in Boston. Two years after that event, the exports of cotton amounted to 93,900,-000 lbs. War with Great Britain, soon afterward, checked both our exports and her manufacture of the article; but the year 1817, memorable in this connection, from its being the date of the organization of the Colonization Society, found our exports augmented to 95,-660,000 lbs., and her consumption enlarged to 126,240,000 lbs. Carding and spinning machinery had now reached a good degree of perfection, and the power-loom was brought into general use in England, and was also introduced into the United States. Steamboats, too, were coming into use, in both countries; and great activity prevailed in commerce, manufactures, and the cultivation of cotton.

But how fared it with the free colored people, during all this time? We must revert to the days of the Pennsylvania Abolition Society, to obtain a true answer to this question.

With freedom to the slave, came anxieties among the whites as to the results. Nine years after Pennsylvania and Massachusetts had taken the lead in the trial of emancipation, FRANKLIN issued an Appeal for aid to enable his Society to form a plan for the promotion of industry, intelligence, and morality among the free blacks; and he zealously urged the measure on public attention, as essential to their well-being, and indispensable to the safety of society. He expressed his belief, that such is the debasing influence of Slavery on human nature, that its very extirpation, if not performed with care, may sometimes open a source of serious evils; and that so far as emancipation should be promoted by the Society, it was a duty incumbent on its members to instruct, to advise, to qualify those restored to freedom, for the exercise and enjoyment of civil liberty.

How far Franklin's influence failed to promote the humane object he had in view, may be inferred from the fact, that forty-seven years after Pennsylvania struck off the shackles from her slaves, and thirty-eight after he issued his Appeal, one-third of the convicts in her penitentiary were colored men, while few of the other free States were more fortunate; and some of them even worse—one-half of New Jersey's convicts being colored men.*

Had the freedmen, in the Northern States, improved their privileges; had they established a reputation for industry, integrity, and virtue, far other consequences would have followed their emancipation. Their advancement in moral character, would have put to shame the advocate for the perpetuation of Slavery. Indeed, there could have been no plausible argument found for its continuance. No regular exports of cotton, no cultivation of cane sugar (to give a profitable character to Slave labor), had any existence when JAY and FRANKLIN commenced their labors, and when Congress took its first step for the suppression of the Slave trade.

Unfortunately, the free colored people persevered in their evil habits. This not only served to fix their own social and political condition on the level of the slave, but it reacted with fearful effect upon their brethren remaining in bondage. Their refusing to listen to the counsel of

*See Boston Prison Discipline Society's Reports, 1826–7.

the philanthropists, who urged them to forsake their indolence and vice, and their frequent violations of the laws, more than all things else, put a check to the tendencies, in public sentiment, toward general emancipation. The failure of FRANKLIN'S plan for their elevation, confirmed the popular belief, that such an undertaking was impracticable; and the whole African race, freedmen as well as slaves, were viewed as an intolerable burden—such as the imports of foreign paupers are now considered. Thus the free colored people themselves, ruthlessly threw the car of emancipation from the track, and tore up the rails upon which, alone, it could move.

The opinion that the African race would become a growing burden had its origin long before the Revolution, and led the colonists to oppose the introduction of slaves; but, failing in this, through the opposition of England, as soon as they threw off the foreign yoke, many of the States at once crushed the system—among the first acts of sovereignty by Virginia, being the prohibition of the Slave trade. In the determination to suppress this traffic all the States united—but in emancipation their policy differed. It was found easier to manage the slaves than the free blacks—at least it was claimed to be so—and, for this reason, the Slave States, not long after the others had completed their work of manumission, proceeded to enact laws prohibiting emancipations, except on condition that the persons liberated should be removed. The newly organized Free States, too, taking alarm at this, and dreading the influx of the free colored people, adopted measures to prevent the ingress of this proscribed and helpless race.

These movements, so distressing to the reflecting colored man, be it remembered, were not the effect of the action of Colonizationists, but took place, mostly, long before the organization of the American Colonization Society; and, at its first annual meeting, the importance and humanity of colonization was strongly urged, on the very ground that the Slave States, as soon as they should find that the persons liberated could be sent to Africa, would relax their laws against emancipation.

The slow progress made by the great body of the free blacks in the North, or the absence, rather, of any evidences of improvement in industry, intelligence, and morality, gave rise to the notion, that before they could be elevated to an equality with the whites, Slavery must be wholly abolished throughout the Union. The constant ingress of liberated slaves from the South, to commingle with the free colored people of the North, tended to perpetuate the low moral standard originally existing among the blacks; and universal emancipation was believed to be indispensable to the elevation of the race. Those who adopted this view, seem to have overlooked the fact, that the Africans, of savage origin, could not be elevated at once to an equality with the American people, by the mere force of legal enactments. More than this was needed, for their elevation, as all are now, reluctantly, compelled to acknowledge.

The Slave States adopted opinions, as to the negro character, opposite to those of the Free States, and would not risk the experiment of emancipation. They said, if the Free States feel themselves burdened by the few persons, of African descent, they have freed, and find it impracticable to educate and elevate, how much greater would be the evil the Slave States must bring upon themselves, by letting loose a population nearly twelve times as numerous. Such an act, they argued, would be suicidal—it would crush out all progress in civilization; or, in the effort to elevate the negro with the white man—allowing him equal freedom of action—would be to make the more energetic Anglo-Saxon the slave of the indolent African. Such a task, onerous in the highest degree, they could not, and would not undertake—such an experiment, on their social system, they dared not hazard.

Another question, "How shall the Slave trade be suppressed? began to be agitated near the close of the last century. The moral desolation existing in Africa, was without a parallel among the nations of the earth. When the last of our Northern States had freed its slaves, not a single Christian Church had been successfully established in Africa, and the Slave trade was still legalized to the citizens of every Christian nation. Even its subsequent prohibition, by the United States and England, had no tendency to check the traffic, nor ameliorate the condition of the African. The other European powers, having now the monopoly of the trade, continued to prosecute it with a vigor it never felt before. The institution of Slavery, while lessened in the United States, where it had not yet been made profitable, was rapidly acquiring an unprecedented enlargement in Cuba and Brazil, where its profitable character had been more fully realized. How shall the Slave trade be annihilated, Slavery extension prevented, and Africa receive a Christian civilization? were questions that agitated the bosom of many a philanthropist, long after WILBERFORCE had achieved his triumphs.

At this period in the history of Africa, and of public sentiment on Slavery, the American Colonization Society was organized. It began its labors when the eye of the statesman, the philanthropist, and the Christian, could discover no other plan of overcoming the moral desolation, the universal oppression, of the colored race, than by restoring the most enlightened of their number to Africa itself. Emancipation, by States, had been at an end for a dozen of years. The improvement of the free colored people, in the presence of the slave, was considered impracticable. Slave labor had become so profitable, as to leave little ground to expect general emancipation, even though all other objections had been removed. The Slave trade had increased twenty-five per cent. during the preceding ten years. Slavery was rapidly extending itself in the tropics, and could not be arrested but by the suppression of the Slave trade. The foothold of the Christian missionary was yet so precarious in Africa, as to leave it doubtful whether he could sustain his position.

The Colonization of the free colored people in Africa, under the teachings of the Christian men who were prepared to accompany them, it was believed, would as fully meet all the conditions of the race, as was possible in the then existing state of the world. It would separate those who should emigrate from all further contact with Slavery, and from its contaminating influences; it would relax the laws of the Slave States against emancipation, and lead to the more frequent liberation of slaves; it would stimulate and encourage the colored people remaining here, to engage in efforts for their own elevation; it would establish free republics along the coast of Africa, and drive away the Slave-trader; it would prevent the extension of Slavery, by means of the Slave trade, in tropical America; it would introduce civilization and Christianity among the people of Africa, and overturn their barbarism and bloody superstitions; and, if successful, it would react upon slavery at home, by pointing out to the States and General Government, a mode by which they might free themselves from the whole African race.

The Society had thus undertaken as great an amount of work as it could perform. The field was broad enough, truly, for an association that hoped to obtain an income of but five to ten thousand dollars a year, and realized annually an average of only $3,276 during the first six years of its existence. It did not, therefore, include the destruction of American Slavery among the objects it labored to accomplish. That subject had been fully discussed; the ablest men in the nation had labored for its overthrow; more than half the original States of the Union had emancipated their slaves; the advantages of freedom to the colored man had been tested: the results had not been as favorable as anticipated; the public sentiment of the country was adverse to an increase of the free colored population; the few of their number who had risen to respectability and affluence, were too widely separated to act in concert in promoting measures for the general good; and, until better results should follow the liberation of slaves, further emancipations, by the States, were not to be expected. The friends of the Colonization Society, therefore, while affording every encouragement to emancipation by individuals, refused to agitate the question of the general abolition of Slavery. Nor did they thrust aside any other scheme of benevolence in behalf of the African race. Forty years had elapsed from the commencement of emancipation in the country, and thirty from the date of FRANKLIN'S APPEAL, before the Society sent off its first emigrants. At that date, no extended plans were in existence, promising relief to the free colored man. A period of lethargy among the benevolent, had succeeded the State emancipations, as a consequence of the indifference of the free colored people, as a class, to their degraded condition. The public sentiment of the country, therefore, was fully prepared to adopt Colonization as the best means, or rather, as the only means for accomplishing anything for them, or for the African race. Indeed, so general was the sentiment in favor of Colonization, somewhere beyond the limits of the United States, that those who disliked Africa, commenced a scheme of emigration to Hayti, and prosecuted it, until 8,000 free colored persons were removed to that island—a number nearly equaling the whole emigration to Liberia up to 1850. Haytian emigration, however, proved a most disastrous experiment.

But the general acquiescence in the objects of

the Colonization Society did not long continue. The exports of cotton from the South were then rapidly on the increase. Slave labor had become profitable, and slaves, in the cotton-growing States, were no longer considered a burden. Seven years after the first emigrants reached Liberia, the South exported 294,310,115 pounds of cotton; and, the year following, the total cotton crop reached 325,000,000 pounds. But a great depression in prices* was now upon the planters, and alarmed them for their safety. They had decided against emancipation, and now to have their slaves rendered valueless, was an evil they were determined to avert.

At this juncture, a warfare against Colonization was commenced at the South, and it was pronounced an Abolition scheme in disguise. In defending itself, the Society re-asserted its principles of neutrality in relation to Slavery, and that it had only in view the Colonization of the free colored people. In the heat of the contest, the South were reminded of their for-

* See Table I, Appendix.

mer sentiments in relation to the whole colored population, and that Colonization merely proposed removing one division of a people they had pronounced a public burden.*

The Emancipationists at the North had only lent their aid to Colonization, in the hope that it would prove an able auxiliary to Abolition; but when the Society declared its unalterable purpose to adhere to its original position of neutrality, they withdrew their support, and

---

* The sentiment of the Colonization Society, was expressed in the following resolution, embraced in its Annual Report of 1826:

"*Resolved*, That the Society disclaims, in the most unqualified terms, the designs attributed to it, of interfering, on the one hand, with the legal rights and obligations of Slavery; and, on the other, of perpetuating its existence within the limits of the country."

On another occasion MR. CLAY, in behalf of the Society, defined its position thus:

"It protested, from the commencement, and throughout all its progress, and it now protests, tlat it entertains no purpose, on its own authority, or by its own means, to attempt emancipation, partial or general; that it knows the General Government has no constitutional power to achieve such an object; that it believes that the States, and the States only, which tolerate Slavery, can accomplish the work of emancipation; and that it ought to be left to them exclusively, absolutely, and voluntarily, to decide the question."—*Tenth Annual Report, p.* 14, 1828.

commenced hostilities against it. "The Anti-slavery Society," said a distinguished Abolitionist, "began with a declaration of war against the Colonization Society."* This feeling of hostility was greatly increased by the action of the Abolitionists of England. The doctrine of "Immediate, not Gradual Abolition," was announced by them, as their creed; and the Anti-slavery men of the United States adopted it as the basis of their action. Its success in the English Parliament, in procuring the passage of the Act for West India Emancipation, in 1833, gave a great impulse to the Abolition cause in the United States.

In 1832, Mr. LLOYD GARRISON declared hostilities against the Colonization Society; in 1834, JAMES G. BIRNEY followed his example; and, in 1836, GERRITT SMITH also abandoned the cause. The North everywhere resounded with the cry of "Immediate Abolition;" and, in 1837, the Abolitionists numbered 1,015 societies; had seventy agents under commission,

---

* GERRITT SMITH, 1835.

and an income, for the year, of $36,000.* The Colonization Society, on the other hand, was greatly embarrassed. Its income, in 1838, was reduced to $10,900; it was deeply in debt; the parent Society did not send a single emigrant, that year, to Liberia; and its enemies pronounced it bankrupt and dead.†

But did the Abolitionists succeed in forcing Emancipation upon the South, when they had thus rendered Colonization powerless? Did the fetters fall from the slave at their bidding? Did fire from heaven descend, and consume the Slaveholder at their invocation? No such thing! They had not touched the true cause of the extension of Slavery. They had not discovered the secret of its power; and, therefore, its locks remained unshorn, its strength unabated. The institution progressed as triumph-

* LUNDY'S LIFE.

† On the floor of an Ecclesiastical Assembly, one minister pronounced Colonization a "dead horse;" while another claimed that his "old mare was giving freedom to more slaves, by trotting off with them to Canada, than the Colonization Society was sending of emigrants to Liberia.

antly as if no opposition existed. The planters were progressing steadily, in securing to themselves the monopoly of the cotton markets of Europe, and in extending the area of Slavery at home. In the same year that GERRITT SMITH declared for Abolition, the title of the Indians to fifty-five millions of acres of land, in the Slave States, was extinguished, and the tribes removed. The year that Colonization was depressed to the lowest point, the exports of cotton, from the United States, amounted to 595,952,297 pounds, and the consumption of the article in England, to 477,206,108 pounds.

When Mr. BIRNEY seceded from Colonization, he encouraged his new allies with the hope, that West India free labor would render our slave labor less profitable, and emancipation, as a consequence, be more easily effected. How stood this matter six years afterward? This will be best understood by contrast. In 1800, the West Indies exported 17,000,000 lbs. of cotton, and the United States, 17,789,803 lbs. They were then about equally productive in that article. In 1840, the West India exports had dwindled down to 427,529 lbs., while those of the United States had increased to 743,941,061 lbs.

And what was England doing all this while? Having lost her supplies from the West Indies, she was quietly spinning away at American Slave-labor cotton; and, to ease the public conscience of the kingdom, was loudly talking of a Free-labor supply of the commodity from the banks of the Niger! But the expedition up that river failed, and 1845 found her manufacturing 626,496,000 lbs. of cotton, mostly the product of American slaves! The strength of American Slavery at that moment, may be inferred from the fact, that we exported, that year, 872,905,996 lbs. of cotton, and our production of cane sugar had reached over 200,000,000 lbs.; while, to make room for Slavery extension, we were busied in the annexation of Texas, and in preparations for the consequent war with Mexico!

But Abolitionists themselves, some time before this, had mostly become convinced of the feeble character of their efforts against Slavery, and allowed politicians to enlist them in a political crusade, as the last hope of arresting the progress of the system. The cry of "Immediate Abolition" died away; reliance upon moral means was mainly abandoned; and the limitation of the institution, geographically, became the chief object of effort. The results of more than a dozen years of political action are before the public, and what has it accomplished! We are not now concerned in the inquiry of how far the strategy of politicians succeeded in making the votes of Abolitionists subservient to Slavery extension. That they did so, in at least one prominent case, will never be denied by any candid man. All we intend to say, is, that the cotton planters, instead of being crippled in their operations, were able, in the year ending the last of June, 1853, to export 1,111,570,370 lbs. of cotton, beside supplying over 400,000,000 lbs. for home consumption; and that England, the year ending the last of January 1853, consumed the unprecedented quantity of 817,998,048 lbs. of that staple. The year 1854, instead of finding Slavery perishing under the blows it had received, has witnessed the destruction of all the old barriers to its extension, and beholds it expanded widely enough for the profitable employment of the slave population, with all its natural increase, for a hundred years to come!!

If *political action* against Slavery has been thus disastrously unfortunate, how is it with *Anti-Slavery action*, at large, as to its efficiency at this moment? On this point hear the testimony of a correspondent of FREDERICK DOUGLASS' *Paper, January* 26, 1855:

"How gloriously did the Anti-Slavery cause arise * * in 1833–4! And now what is it, in our agency! * * What is it through the errors or crimes of its advocates variously—probably quite as much as through the brazen, gross, and licentious wickedness of its enemies. Alas! what is it but a mutilated, feeble, discordant and half-expiring instrument, at which Satan and his children, legally and illegally scoff! Of it I despair."

Such are the crowning results of both political and Anti-Slavery action, for the overthrow of Slavery! Such are the demonstrations of their utter impotency as a means of relief to the bond and free of the colored race!

Surely, then, it is *a time for work*, in some other mode, than that hitherto adopted. Surely, too, it is a time for the American people to rebuke that class of politicians, both North and South, whose only capital consists in keeping up a fruitless warfare upon the subject of Slavery—fruitless, truly, to the colored man — nay! abundant in fruits; but to him, "their vine is of the vine of Sodom, and of the fields of Gomorrah; their grapes of gall, their clusters are bitter; their wine is the poison of dragons, and the cruel venom of asps."*

The application of this language, to the case under consideration, will be fully justified, when the facts are presented in the remaining pages of this work.

2. The present relations of American Slavery to the Industrial interests of our own country; to the demands of Commerce, and to the present Political crisis.

The institution of Slavery, at this moment,

* Deuteronomy xxxii, 32, 33.

gives indications of a vitality that was never anticipated by its friends or foes. Its enemies often supposed it about ready to expire, from the wounds they had inflicted, when in truth it had taken two steps in advance; while they had taken twice the number in an opposite direction. In each successive conflict, its assailants have been weakened, while its dominion has been extended.

This has arisen from causes too generally overlooked. Slavery is not an isolated system, but is so mingled with the business of the world, that it derives facilities from the most innocent transactions. Capital and labor, in Europe and America, are largely employed in the manufacture of cotton. These goods, to a great extent, may be seen freighting every vessel, from Christian nations, that traverses the seas of the globe; and filling the warehouses and shelves of the merchants, over two-thirds of the world. By the industry, skill, and enterprise, employed in the manufacture of cotton, mankind are better clothed; their comfort better promoted; general industry more highly stimulated; commerce more widely extended; and civilization more rapidly advanced, than in any preceding age.

To the superficial observer, all the agencies, based upon the manufacture and sale of cotton, seem to be legitimately engaged in promoting human happiness; and he, doubtless, feels like invoking Heaven's choicest blessings upon them. When he sees the stockholders in the cotton corporations receiving their dividends, the operatives their wages, the merchants their profits, and civilized people everywhere clothed comfortably in cottons, he can not refrain from explaining: "The lines have fallen unto them in pleasant places; yea, they have a goodly heritage!"

But turn a moment to the source whence the raw cotton, the basis of these operations, is obtained, and observe the aspect of things in that direction. When the statistics on the subject are examined, it appears that nearly all the cotton consumed in the Christian world, is the product of the Slave labor of the United States.* It is this monopoly that has given Slavery its commercial value; and, while this monopoly is

* See Appendix, Table I.

retained, the institution will continue to extend itself wherever it can find room to spread. He who looks for any other result, must expect that nations, which, for centuries, have waged war to extend their commerce, will now abandon their means of aggrandizement, and bankrupt themselves, to force the abolition of American Slavery!

This is not all. The economical value of Slavery as an agency for supplying the means of extending manufactures and commerce, has long been understood by statesmen. The discovery of the power of steam, and the inventions in machinery, for preparing and manufacturing cotton, revealed the important fact, that a single Island, having the monopoly secured to itself, could supply the world with clothing. *Great Britain attempted to gain this monopoly;* and, to prevent other countries from rivaling her, she long prohibited all emigration of skillful mechanics from the kingdom, as well as all exports of machinery. As country after country was opened to her commerce, the markets for her manufactures were extended, and the demand for the raw material increased. The benefits of this enlarged commerce of the world, were not confined to a single nation, but mutually enjoyed by all. As each had products to sell, peculiar to itself, the advantages often gained by one, were no detriment to the others. The principal articles demanded by this increasing commerce, have been coffee, sugar, and cotton—in the production of which Slave labor has greatly predominated. Since the enlargement of manufactures, cotton has entered more extensively into commerce than coffee and sugar, though the demand for all three has advanced with the greatest rapidity. England could only become a great commercial nation, through the agency of her manufactures. She was the best supplied, of all the nations, with the necessary capital, skill, labor, and fuel, to extend her commerce by this means. But, for the raw material, to supply her manufactories, she was dependent upon other countries. The planters of the United States were the most favorably situated for the cultivation of cotton, and attempted *to monopolize the markets for that staple.* This led to a fusion of interests between them and the manufacturers of Great Britain; and to the invention of notions, in political economy, that would, so far as adopted, promote the interests of this coalition. With the advantages possessed by the English manufacturers, "Free Trade" would render all other nations subservient to their interests; and, so far as their operations should be increased, just so far would the demand for American cotton be extended. The details of the success of the parties to this combination, and the opposition they have had to encounter, are left to be noticed more fully hereafter. To the cotton planters, the copartnership has been eminently advantageous.

How far the other agricultural interests of the United States are promoted, by extending the cultivation of cotton, may be inferred from the Census returns of 1850, and the Congressional Reports on Commerce and Navigation, for 1854.* Cotton and tobacco, only, are largely exported. The production of sugar does not yet equal our consumption of the article, and we import, chiefly from Slave-labor countries,

* See Appendix, Table II.

445,445,680 lbs. to make up the deficiency.* But of cotton and tobacco, we export more than *two-thirds* of the amount produced; while of other products, of the agriculturists, less than the *one-forty-sixth* part is exported. Foreign nations, generally, can grow their provisions, but can not grow their tobacco and cotton. Our surplus provisions, not exported, go to the villages, towns, and cities, to feed the mechanics, manufacturers, merchants, professional men, and others; or to the cotton and sugar districts of the South, to feed the planters and their slaves. The increase of mechanics and manufacturers at the North, and the expansion of Slavery at the South, therefore, augment the markets for provisions, and promote the prosperity of the farmer. As the mechanical population increases, the implements of husbandry, and articles of furniture, are multiplied, so that both farmer and planter can be supplied with them on easier terms. As foreign nations open their markets to cotton fabrics, increased demands, for the raw material, are made. As new grazing and grain-

* Table III.

growing States are developed, and teem with their surplus productions, the mechanic is benefited, and the planter, relieved from food-raising, can employ his slaves more extensively upon cotton. It is thus that our exports are increased; our foreign commerce advanced; the home markets of the mechanic and farmer extended, and the wealth of the nation promoted. It is thus, also, that the Free labor of the country finds remunerating markets for its products—though at the expense of serving as an efficient auxiliary in the extension of Slavery!

But more. So speedily are new grain-growing States springing up; so vast is the territory owned by the United States, ready for settlement; and so enormous will soon be the amount of products demanding profitable markets, that the national government has been seeking new outlets for them, upon our own continent, to which, alone, they can be advantageously transported. That such outlets, when our vast possessions, Westward, are brought under cultivation, will be an imperious necessity, is known to every statesman. The farmers of these new

States, after the example of those of the older sections of the country, will demand a market for their products. This can be furnished, only, by the extension of Slavery; by the acquisition of more tropical territory; by opening the ports of Brazil, and other South American countries, to the admission of our provisions; or by a vast enlargement of domestic manufactures, to the exclusion of foreign goods from the country. Look at this question as it now stands, and then judge of what it must be twenty years hence. The class of products under consideration, in the whole country, in 1853, were valued at $1,551,-176,490; of which there were exported to foreign countries, to the value of only $33,809,-126.* The planter will not assent to any check upon the foreign imports of the country, for the benefit of the farmer. This demands the adoption of vigorous measures to secure a market for his products by some of the other modes stated. Hence, the orders of our Executive, in 1851, for the exploration of the valley of the Amazon; the efforts, in 1854, to obtain a treaty

* See Appendix, Table II.

with Brazil for the free navigation of that immense river; the negotiations for a military foothold in St. Domingo, and the determination to acquire Cuba. But we must not anticipate topics to be considered at a later point in our discussion.

Antecedent to all these movements, Great Britain had foreseen the coming increased demand for tropical products. Indeed, her West Indian policy, of a few years previous, had hastened the crisis; and, to repair her injuries, and meet the general outcry for cotton, she made the most vigorous efforts to promote its cultivation in her own tropical possessions. The motives prompting her to this policy, need not be referred to here, as they will be noticed hereafter. The Hon. GEORGE THOMPSON, it will be remembered, when urging the increase of cotton cultivation in the East Indies, declared that the scheme must succeed, and that, soon, all Slave-labor cotton would be repudiated by the British manufacturers. Mr. GARRISON indorsed the measure, and expressed his belief that, with its success, the American Slave system must inevitably perish, from starvation! But England's efforts signally failed, and the golden apple, fully ripened, dropped into the lap of our cotton planters. The year that heard THOMPSON'S pompous predictions,* witnessed the consumption of but 445,744,000 lbs. of cotton, by England, while, fourteen years later, she used 817,998,048 lbs., nearly 700,000,000 lbs. of which were obtained from America!

That we have not overstated her dependence upon our Slave labor for cotton, is a fact of world-wide notoriety. Blackwood's Magazine, January, 1853, in referring to the cultivation of the article, by the United States, says:

"With its increased growth has sprung up that mercantile navy, which now waves its stripes and stars over every sea, and that foreign influence, which has placed the internal peace—we may say the subsistence of millions, in every manufacturing country in Europe—within the power of an oligarchy of planters."

In reference to the same subject, the London Economist quotes as follows:

"Let any great social or physical convulsion

* 1839.

visit the United States, and England would feel the shock from Land's End to John O'Groats. The lives of nearly two millions of our countrymen are dependent upon the cotton crops of America; their destiny may be said, without any kind of hyperbole, to hang upon a thread. Should any dire calamity befall the land of cotton, a thousand of our merchant ships would rot idly in dock; ten thousand mills must stop their busy looms; two thousand thousand mouths would starve, for lack of food to feed them."

A more definite statement of England's indebtedness to cotton, is given by McCulloch; who shows, that as far back as 1832, her exports of cotton fabrics were equal in value to about *two-thirds* of all the woven fabrics exported from the empire. The same state of things, nearly, existed in 1849, when the cotton fabrics exported were valued at about $140,000,000, while all the other woven fabrics exported did not quite reach to the value of $68,000,000.*

* London Economist, 1850.

There was a time when American Slave labor sustained no such relations to the manufactures and commerce of the world as it now so firmly holds; and when, by the adoption of proper measures, on the part of the Free colored people and their friends, the emancipation of the slaves, in all the States, might have been effected. But that period has passed forever away, and causes, unforeseen, have come into operation, which are too powerful to be overcome by any agencies that have since been employed. What Divine Providence may have in store for the future, we know not; but, at present, the institution of Slavery is sustained by numberless pillars, too massive for human power and wisdom to overthrow.

Take another view of this subject. To say nothing now of the tobacco, rice, and sugar, which are the products of our slave labor, we exported raw cotton to the value of $109,456,404 in 1853. Its destination was, to Great Britain, 768,596,498 lbs.; to the Continent of Europe, 335,271,434 lbs.; to countries on our own Continent, 7,702,438 lbs.; making the total exports, 1,111,570,370 lbs. The entire crop of that year being 1,600,000,000 lbs., gives, for home consumption, 488,429,630 lbs. Of this, there was manufactured into cotton fabrics to the value of $61,869,274;* of which there was retained, for home markets, to the value of $53,100,290. Our imports of cotton fabrics from Europe, in 1853, for consumption, amounted in value to $26,477,950: thus making our cottons, foreign and domestic, for that year, cost us $79,578,240.

This, now, is what becomes of our *cotton;* this is the way in which it so largely constitutes the basis of commerce and trade; and this is the nature of the relations existing between the Slavery of the United States and the material interests of the world.

But have the United States no other great leading interests, except those which are involved in the production of cotton? Certainly,

* This estimate is probably too low, being taken from the census of 1850. The exports of cottons for 1850 were, $4,734,424, and for 1853, $8,768,894; having nearly doubled in four years.

they have. Here is a great field for the growth of *provisions.* In ordinary years, exclusive of tobacco and cotton, our agricultural property, when added to the domestic animals and their products, amounts in value to $1,551,176,490. Of this, there is exported only to the value of $33,809,126; which leaves for home consumption and use, a remainder to the value of $1,517,367,364.* The portions of the property represented by this immense sum of money, which pass from the hands of the agriculturists, are distributed throughout the Union, for the support of the day-laborers, sailors, mechanics, manufacturers, traders, merchants, professional men, planters, and the slave population. This is what becomes of the *provisions.*

Beside this annual consumption of provisions, most of which is the product of *free labor,* the people of the United States use a vast amount of *groceries,* which are mainly of *Slave labor* origin. Boundless as is the influence of cotton,

* See Table II, Appendix.

in stimulating Slavery extension, that of the cultivation of groceries falls but little short of it; the chief difference being, that they do not receive such an increased value under the hand of manufacturers. The cultivation of coffee, in Brazil, employs as great a number of slaves as that of cotton in the United States.

But, to comprehend fully our indebtedness to slave labor for groceries, we must descend to particulars. Our imports of coffee, tobacco, sugar, and molasses, for 1853, amounted in value to $38,479,000; of which the hand of the slave, in Brazil and Cuba mainly, supplied to the value of $34,451,000.[*] This shows the extent to which we are sustaining *foreign Slavery*, by the consumption of these four products. But this is not our whole indebtedness to Slavery for groceries. Of the domestic grown tobacco, valued at $19,975,000, of which we retain nearly one-half, the Slave States produce to the value of $16,787,000; of domestic rice, the product of the South, we consume to the

---

[*] See Table III, Appendix.

value of $7,092,000; of domestic Slave grown sugar and molasses, we take, for home consumption, to the value of $34,779,000; making our grocery account with *domestic Slavery*, foot up the sum of $50,449,000. Our whole indebtedness, then, to Slavery, foreign and domestic, for these four commodities, after deducting two millions of re-exports, amounts to $82,607,000.

By adding the value of the foreign and domestic cotton fabrics, consumed annually in the United States, to the yearly cost of the groceries which the country uses, our total indebtedness, for articles of Slave labor origin, will be found swelling up to the enormous sum of $162,185,240.

We have now seen the channels through which our *cotton* passes off into the great sea of commerce, to furnish the world its clothing. We have seen the origin and value of our *provisions*, and to whom they are sold. We have seen the sources whence our *groceries* are derived, and the millions of money they cost. To ascertain how far these several interests are sustained by one another, will be to determine how far any one of them becomes an element of expansion to the others. To decide a question of this nature, with precision, is impracticable. The statistics are not attainable. It may be illustrated, however, in various ways, so as to obtain a conclusion proximately accurate. Suppose, for example, that the supplies of food from the North were cut off, the manufactories left in their present condition, and the planters forced to raise their provisions and draught animals: in such circumstances, the export of cotton must cease, as the lands of these States could not be made to yield more than would subsist their own population, and supply the cotton demanded by the Northern States. Now, if this be true of the agricultural resources of the cotton States—and it is believed to be the full extent of their capacity—then the surplus of cotton, to the value of more than a hundred millions of dollars, now annually sent abroad, stands as the representative of the yearly supplies which the cotton planters receive from the farmers north of the cotton line. This, therefore, as will afterward more fully appear, may be taken as the probable extent to which the supplies from the North serve as an element of Slavery expansion, in the article of cotton alone.

But this subject demands a still closer scrutiny, as to its past connections with national politics, in order that the causes of the failure of Abolitionism to arrest the progress of Slavery, as well as the present relations of the institution to the politics of the country, may fully appear.

Slave labor has seldom been made profitable where it has been wholly employed in grazing and grain growing; but it becomes remunerative in proportion as the planters can devote their attention to cotton, sugar, rice, or tobacco. To render Southern Slavery profitable in the highest degree, therefore, the slaves must be employed upon some one of these articles, and be sustained by a supply of food and draught animals from Northern agriculturists; and, before the planter's supplies are complete, to these must be added cotton gins, implements of husbandry, furniture, and tools, from Northern mechanics. This is a point of the utmost moment, and must be considered more at length.

It has long been a vital question to the success of the Slaveholder, to know how he could render the labor of his slaves the most profitable. The grain growing States had to emancipate their slaves, to rid themselves of a profitless system. The cotton growing States, ever after the invention of the cotton gin, had found the production of that staple, highly remunerative. The logical conclusion, from these different results, was, that the less provisions, and the more cotton grown by the planter, the greater would be his profits. *Markets* for the surplus products of the farmer of the North, were equally as important to him as the supply of *provisions* was to the planter. But the planter, to be eminently successful, must purchase his supplies, at the lowest possible prices; while the farmer, to secure his prosperity, must sell his products at the highest possible rates. Few, indeed, can be so ill informed, as not to know, that these two topics, for many years, were involved in the "Free Trade" and "Protective Tariff" doctrines, and afforded the *materiel* of the political contests between the North and the South—between free labor and slave labor. A very brief notice of the history of that controversy, will demonstrate the truth of this assertion.

The attempt of the agricultural States, thirty years since, to establish the protective policy, and promote "Domestic Manufactures," was a struggle to create such a division of labor, as would afford a "Home Market" for their products, no longer in demand abroad. The first decisive action on the question, by Congress, was in 1824; when the distress in these States, and the measures proposed for their relief, by national legislation, were discussed on the passage of the "Tariff Bill" of that year. The ablest men in the nation were engaged in the controversy. As provisions are the most important item on the one hand, and cotton on the other, we shall use these two terms as the representatives of the two classes of products, belonging, respectively, to Free labor and to Slave labor.

Mr. CLAY, in the course of the debate, said: "What, again, I would ask, is the CAUSE of the unhappy condition of our country, which I have fairly depicted? It is to be found in the fact that, during almost the whole existence of this government, we have shaped our industry, our navigation, and our commerce, in reference to an extraordinary war in Europe, and to foreign markets which no longer exist; in the fact that we have depended too much on foreign sources of supply, and excited too little the native; in the fact that, while we have cultivated, with assiduous care, our foreign resources, we have suffered those at home to wither, in a state of neglect and abandonment. The consequence of the termination of the war of Europe, has been the resumption of European commerce, European navigation, and the extension of European agriculture, in all its branches. Europe, therefore, has no longer occasion for anything like the same extent as that which she had during her wars, for American commerce, American navigation, the produce of American industry. Europe in commotion, and convulsed throughout all her members, is to America no longer the same Europe as she is now, tranquil, and watching with the most vigilant attention, all her own peculiar interests, without regard to their operation on us. The effect of this altered state of Europe upon us, has been to circumscribe the employment of our marine, and greatly to reduce the value of the produce of our territorial labor. * * The greatest want of civilized society is a market for the sale and exchange of the surplus of the products of the labor of its members. This market may exist at home or abroad, or both, but it must exist somewhere, if society prospers; and, wherever it does exist, it should be competent to the absorption of the entire surplus production. It is most desirable that there should be both a home and a foreign market. But with respect to their relative superiority, I can not entertain a doubt. The home market is first in order, and paramount in importance. The object of the bill under consideration, is to create this home market, and to lay the foundations of a genuine American policy. It is opposed; and it is incumbent on the partisans of the foreign policy (terms which I shall use without any invidious intent) to demonstrate that the foreign market is an adequate vent for the surplus produce of our labor. But is it so? 1. Foreign nations can not, if they would, take our surplus produce. * * 2. If they could, they would not. * * We have seen, I think, the causes of the distress of the country. We have seen that an exclusive dependence upon the foreign market must lead to a still severer distress, to impoverishment, to ruin. We must, then, change somewhat our course. We must give a new direction to some portion of our industry. We must speedily adopt a genuine American policy. Still cherishing a foreign market, let us create also a home market, to give further scope to the consumption of the produce of American industry. Let us counteract the policy of foreigners, and withdraw the support which we now give to their industry, and stimulate that of our own country. * * The creation of a home market is not only necessary to procure for our agriculture a just reward of its labors, but it is indispensable to obtain a supply of our necessary wants. If we can not sell, we can not buy. That portion of our population (and we have seen that it is not less than four-fifths) which makes comparatively nothing that foreigners will buy, has nothing to make purchases with from foreigners. It is in vain that we are told of the amount of our exports, supplied by the planting interest. They may enable the planting interest to supply all its wants; but they bring no ability to the interests not planting, unless, which can not be pretended, the planting interest was an adequate vent for the surplus produce of all the labor of all other interests. * * But this home market, highly desirable as it is, can only be created and cherished by the PROTECTION of our own legislation against the inevitable prostration of our industry, which must ensue from the action of FOREIGN policy and legislation. * * The sole object of the tariff is to tax the produce of foreign industry, with the view of promoting American industry. * * But it is said by the honorable gentleman from Virginia, that the South, owing to the character of a certain portion of its population, can not engage in the business of manufacturing. * * The circumstances of its degradation unfits it for the manufacturing arts. The well-being of the other, and the larger part of our population, requires the introduction of those arts.

"What is to be done in this conflict? The gentleman would have us abstain from adopting a policy called for by the interests of the greater and freer part of the population. But is that reasonable? Can it be expected that the interests of the greater part should be made to bend to the condition of the servile part of our population? That, in effect, would be to make us the slaves of slaves. * * I am sure that the patriotism of the South may be exclusively relied upon to reject a policy which should be dictated by considerations altogether connected with that degraded class, to the prejudice of the residue of our population. But, does not a perseverance in the foreign policy, as it now exists, in fact, make all parts of the Union, not planting, tributary to the planting parts? What is the argument? It is, that we must continue freely to receive the produce of foreign industry without regard to the protection of American industry, that a market may be retained for the sale abroad of the produce of the planting portion of the country; and that, if we lessen the consumption, in all parts of America, those which are not planting, as well as the planting sections, of foreign manufactures, we diminish to that extent the foreign market for the planting produce. The existing state of things, indeed, presents a sort of tacit compact between the cotton-grower and the British manufacturer, the stipulations of which are, on the part of the cotton-grower, that the whole of the United States, the other portions as well as the cotton-growing, shall remain open and unrestricted in the consumption of British manufactures; and, on the part of the British manufacturer, that, in consideration thereof, he will continue to purchase the cotton of the South.

Thus, then, we perceive, that the proposed measure, instead of sacrificing the South to the other parts of the Union, seeks only to preserve them from being absolutely sacrificed under the operation of the tacit compact which I have described."

The opposition to the Protective tariff, by the South, arose from two causes—the first openly avowed at the time, and the second clearly deducible from the policy it pursued: the one to secure the foreign market for its cotton, the other to obtain a bountiful supply of Provisions at cheap rates. Cotton was admitted free of duty into foreign countries, and Southern statesmen feared its exclusion, if our government increased the duties on foreign fabrics. The South exported about twice as much of that staple, as was supplied to Europe by all other countries, and there were indications favoring the desire it entertained of monopolizing the foreign markets. The West India planters could not import food, but at such high rates as to make it impracticable to grow cotton at prices low enough to suit the English manufacturer. To purchase cotton cheaply, was essential to the success of his scheme of monopolizing its manufacture, and supplying the world with clothing. The close proximity of the provision and cotton-growing districts, in the United States, gave its planters advantages over all other portions of the world. But they could not monopolize the markets, unless they could obtain a cheap supply of food and clothing for their negroes, and raise their cotton at such reduced prices as to undersell their rivals. A manufacturing population, with its mechanical coadjutors, in the midst of the provision growers, on a scale such as the Protective policy contemplated, it was conceived, would create a permanent market for their products, and enhance the price; whereas, if this manufacturing could be prevented, and a system of free trade adopted, the South would constitute the principal Provision market of the country, and the fertile lands of the North supply the cheap food demanded for its slaves. As the tariff policy, in the outset, contemplated the encouragement of the production of iron, hemp, whisky, and the establishment of woolen manufactories, principally, the South found its interests but slightly identified with the system—the coarser qualities of cottons, only, being manufactured in the country, and, even these, on a diminished scale, as compared with the cotton crops of the South. Cotton, up to the date when this controversy had fairly commenced, had been worth, in the English market, an average price of from 29 7-10 to 48 4-10 cents per lb.* But at this period a wide-spread and ruinous depression, both in the culture and manufacture of the article, occurred—cotton in 1826, having fallen, in England, as low as 11 9-10 to 18 9-10 cents per lb. The home market, then, was too inconsiderable to be of much importance, and there existed little hope of its enlargement to the extent demanded by its increasing cultivation. The planters, therefore, looked abroad to the existing markets, rather than to wait for tardily creating one at home. For success in the foreign markets, they relied, mainly, upon preparing themselves to produce cotton at the reduced prices then prevailing in Europe. All agricultural

* This includes the period from 1806, to 1826, though the decline began a few years before the later date.

products, except cotton, being excluded from foreign markets, the planters found themselves almost the sole *exporters* of the country; and it was to them a source of chagrin, that the North did not, at once, co-operate with them in augmenting the commerce of the nation.

At this point in the history of the controversy, politicians found it an easy matter to produce feelings of the deepest hostility between the opposing parties. The planters were led to believe that the millions of revenue collected off the goods imported, was so much deducted from the value of the cotton that paid for them, either in the diminished price they received abroad, or in the increased price which they paid for the imported articles. To enhance the duties, for the protection of our manufactures, they were persuaded, would be so much of an additional tax upon themselves, for the benefit of the North; and, beside, to give the manufacturer such a monopoly of the home market for his fabrics, would enable him to charge purchasers an excess over the true value of his stuffs, to the whole amount of the duty. By the protective policy, the planters expected to have the cost of both provisions and clothing increased, and their ability to monopolize the foreign markets diminished in a corresponding degree. If they could establish Free trade, it would insure the American market to foreign manufacturers; secure the foreign markets for their leading staple; repress home manufactures; force a larger number of the Northern men into agriculture; multiply the growth, and diminish the price of provisions; feed and clothe their slaves at lower rates; produce their cotton for a third or fourth of former prices; rival all other countries in its cultivation; monopolize the trade in the article throughout the whole of Europe; and build up a commerce and a navy that would make us the ruler of the seas.

But, to understand the sentiments of the South, on the Protective policy, as expressed by its statesmen, we must again quote from the Congressional Debates of 1824:

Mr. HAYNE, of South Carolina, said: "But how, I would seriously ask, is it possible for the home market to supply the place of the foreign market, for our cotton? We supply Great Britain with the raw material, out of which she furnishes the Continent of Europe, nay, the whole world, with cotton goods. Now, suppose our manufactories could make every yard of cloth we consume, that would furnish a home market for more than 20,000,000 lbs. out of the 180,000,000 lbs. of cotton now shipped to Great Britain; leaving on our hands 160,-000,000 lbs., equal to two-thirds of our whole produce. * * * * * * Considering this scheme of promoting certain employments, at the expense of others, as unequal, oppressive, and unjust—viewing prohibition as the *means*, and the destruction of all foreign commerce as the *end* of this policy—I take this occasion to declare, that we shall feel ourselves justified in embracing the very first opportunity of repealing all such laws as may be passed for the promotion of these objects."

Mr. CARTER, of South Carolina, said: "Another danger to which the present measure would expose this country, and one in which the Southern States have a deep and vital interest, would be the risk we incur, by this system of exclusion, of driving Great Britain to countervailing measures, and inducing all other countries, with whom the United States have any considerable trading connections, to resort to measures of retaliation. There are countries possessing vast capacities for the production of rice, of cotton, and of tobacco, to which England might resort to supply herself. She might apply herself to Brazil, Bengal, and Egypt, for her cotton; to South America, as well as to her colonies, for her tobacco; and to China and Turkey for her rice."

Mr. GOVAN, of South Carolina, said: "The effect of this measure on the cotton, rice, and tobacco growing States, will be pernicious in the extreme:—it will exclude them from those markets where they depended almost entirely for a sale of those articles, and force Great Britain to encourage the cottons, (Brazil, Rio Janeiro, and Buenos Ayres,) which, in a short time, can be brought in competition with us. Nothing but the consumption of British goods in this country, received in exchange, can support a command of the cotton market to the Southern planter. It is one thing very certain, she will not come here with her gold and silver to trade with us. And should Great Britain, pursuing the principles of her reciprocal duty act, of last June, lay three or four cents on our cotton, where would, I ask, be our surplus of cotton? It is well known that the United States can not manufacture one-fourth of the cotton that is in it; and should we, by our imprudent legislative enactments, in pursuing to such an extent this restrictive system, force Great Britain to shut her ports against us, it will paralyze the whole trade of the Southern country. This export trade, which composes five-sixths of the export trade of the United States, will be swept entirely from the ocean, and leave but a melancholy wreck behind."

It is necessary, also, to add a few additional extracts, from the speeches of Northern statesmen, during this discussion.

Mr. Martindale, of New York, said: "Does not the agriculture of the country languish, and the laborer stand still, because, beyond the supply of food for his own family, his produce perishes on his hands, or his fields lie waste and fallow; and this because his accustomed market is closed against him? It does, sir. * * * A twenty years' war in Europe, which drew into its vortex all its various nations, made our merchants the carriers of a large portion of the world, and our farmers the feeders of immense belligerent armies. An unexampled activity and increase in our commerce followed—our agriculture extended itself, grew, and flourished. An unprecedented demand gave the farmer an extraordinary price for his produce. * * * Imports kept pace with exports, and consumption with both. * * * Peace came into Europe, and shut out our exports, and found us in war with England, which almost cut off our imports. * * * Now we felt how *comfortable* it was to have plenty of food, but no clothing. * * * Now we felt the imperfect organization of our system. Now we saw the imperfect distribution and classification of labor.

* * * Here is the explanation of our opposite views. It is employment, after all, that we are all in search of. It is a market for our labor and our produce, which we all want, and all contend for. 'Buy foreign goods, that we may import,' say the merchants: it will make a market for importations, and find employment for our ships. Buy English manufactures, say the cotton planters; England will take our cotton in exchange. Thus the merchant and the cotton planter fully appreciate the value of a market when they find their own encroached upon. The farmer and manufacturer claim to participate in the benefits of a market for their labor and produce; and hence this protracted debate and struggle of contending interests. It is a contest for a market between the *cotton grower and merchant* on one side, and the *farmer and manufacturer* on the other. That the manufacturer would furnish this market to the farmer, admits no doubt. The farmer should reciprocate the favor; and government is now called upon to render this market acces-

sible to foreign fabrics for the mutual benefit of both. * * * This, then, is the remedy we propose, sir, for the evils which we suffer. Place the mechanic by the side of the farmer, that the manufacturer who makes our cloth, should make it from *our* farmers' wool, flax, hemp, etc., and be fed by our farmers' provisions. Draw forth our iron from our own mountains, and we shall not drain our country in the purchase of the foreign. * * * We propose, sir, to supply our own wants from our own resources, by the means which God and Nature have placed in our hands. * * * But here is a question of sectional interest, which elicits unfriendly feelings and determined hostility to the bill. * * * The cotton, rice, tobacco, and indigo growers of the Southern States, claim to be deeply affected and injured by this system. * * * Let us inquire if the Southern planter does not demand what, in fact, he denies to others. And now, what does he require? That the North and West should buy — what? Not their cotton, tobacco, etc., for that we do already, to the utmost of our ability to consume, or pay, or vend to others; and that is to an immense amount, greatly exceeding what they purchase of us. But they insist that we should buy English wool, wrought into cloth, that they may pay for it with their cotton; that we should buy Russia iron, that they may sell their cotton; that we should buy Holland gin and linen, that they may sell their tobacco. In fine, that we should not grow wool; and dig and smelt iron of the country; for, if we did, they could not sell their cotton." [On another occasion, he said]: "Gentlemen say they *will* oppose every part of the Bill. They will, therefore, move to strike out every part of it. And, on every such motion, we shall hear repeated, as we have done already, the same objections: that it will ruin trade and commerce; that it will destroy the revenue, and prostrate the navy; that it will enhance the prices of articles of the first necessity, and thus be taxing the poor; and that it will destroy the cotton market, *and stop the further growth of cotton.*"

Mr. BUCHANAN, of Pennsylvania, said: "No

nation can be perfectly independent, which depends upon foreign countries for its supply of iron. It is an article equally necessary in peace and in war. Without a plentiful supply of it, we can not provide for the common defense. Can we so soon have forgotten the lesson which experience taught us during the late war with Great Britain? Our foreign supply was then cut off, and we could not manufacture in sufficient quantities for the increased domestic demand. The price of the article became extravagant, and both the Government and the agriculturist were compelled to pay double the sum for which they might have purchased it, had its manufacture, before that period, been encouraged by proper protecting duties."

Sugar cane, at that period, had become an article of culture in Louisiana, and efforts were made to persuade her planters into the adoption of the Free trade system. It was urged that they could more effectually resist foreign competition, and extend their business, by a cheap supply of food, than by protective duties. But the Louisianians were too wise not to know, that though they would certainly obtain cheap provisions by the destruction of Northern manufactures; still, this would not enable them to compete with the cheaper labor supplied by the Slave-trade to the Cubans.

The West, for many years, gave its undivided support to the manufacturing interests, thereby obtaining a heavy duty on hemp, wool, and foreign distilled spirits: thus securing encouragement to its hemp and wool growers, and the monopoly of the home market for its whisky. The distiller and the manufacturer, under this system, were equally ranked as public benefactors, as each increased the consumption of the surplus products of the farmer. The grain of the West could find no remunerative market, except as fed to domestic animals, for *droving* East and South, or distilled into whisky, which would bear transportation. Take a fact in proof of this assertion. Hon. HENRY BALDWIN, of Pittsburgh, at a public dinner given him by the friends of General Jackson, in Cincinnati, May, 1828, in referring to the want of markets, for the farmers of the West, said, "He was certain, the aggregate of their agricultural produce, finding a market in Europe, would not pay for the *pins* and *needles* they imported."

The markets in the Southwest, now so important, were then quite limited. As the Protective system, coupled with the contemplated internal improvements, if successfully accomplished, would inevitably tend to enhance the price of agricultural products; while the Free trade and anti-internal improvement policy, would as certainly reduce their value; the two systems were long considered so antagonistic, that the success of the one must sound the knell of the other. Indeed, so fully was Ohio impressed with the necessity of promoting manufactures, that all capital, thus employed, was for many years entirely exempt from taxation.

It was in vain that the friends of protection appealed to the fact, that the duties levied on foreign goods did not necessarily enhance their cost to the consumer; that the competition among home manufacturers, and between them and foreigners, had greatly reduced the price of nearly every article properly protected; that foreign manufacturers always had, and always would advance their prices according to our dependence upon them; that domestic competition was the only safety the country had against foreign imposition; that it was necessary we should become our own manufacturers, in a fair degree, to render ourselves independent of other nations in times of war, as well as to guard against the vacillations in foreign legislation; that the South would be vastly the gainer by having the market for its products at its own doors, to avoid the cost of their transit across the Atlantic; that, in the event of the repression or want of proper extension of our manufactures, by the adoption of the free trade system, the imports of foreign goods, to meet the public wants, would soon exceed the ability of the people to pay, and, inevitably, involve the country in bankruptcy.

Southern politicians remained inflexible, and refused to accept any policy except free trade, to the utter abandonment of the principle of protection. Whether they were jealous of the greater prosperity of the North, and desirous to cripple its energies, or whether they were truly fearful of bankrupting the South, we shall not wait to inquire. Justice demands, however, that we should state, that the South was suffering from the stagnation in the cotton trade existing throughout Europe. The planters had been unused to the low prices, for that staple, they were compelled to accept. They had no prospect of an adequate home market for many years to come, and there were indications that they might lose the one they already possessed. The West Indies was still Slave territory, and attempting to recover its early position in the English market. This it had to do, or be forced into emancipation. The powerful Viceroy of Egypt, Mehemet Ali, was endeavoring to compel his subjects to grow cotton on an enlarged scale. The newly organized South American republics were assuming an aspect of commercial consequence, and might commence its cultivation. The East Indies and Brazil were supplying to Great Britain from one-third to one-half of the cotton she was annually manufacturing. The other half, or two-thirds, she might obtain from other sources, and repudiate all traffic with our planters. Southern men, therefore, could not conceive of anything but ruin to themselves, by any considerable advance in duties on foreign imports. They understood the protective policy as contemplating the supply of our country with home manufactured articles, to the exclusion of those of foreign countries. This would confine the planters in the sale of their cotton, mainly to the American market, and leave them in the power of moneyed corporations; which, possessing the ability, might control the prices of their staple, to the irreparable injury of the South. With Slave labor they could not become manufacturers, and must, therefore, remain at the mercy of the North, both as to food and clothing, unless the European markets should be retained. Out of this conviction grew the war upon Corporations; the hostility to the employment of foreign capital in developing the mineral, agricultural, and manufacturing resources of the country; the efforts to destroy the banks and the credit system; the attempts to reduce the currency to gold and silver; the system of collecting the public revenues in coin; the withdrawal of the public moneys from all banks, as a basis of paper circulation; and the sleepless vigilance of the South, in resisting all systems of internal improvements by the General Government. Its statesmen foresaw, that a paper currency would keep up the price of Northern products one or two hundred per cent. above the specie standard; that combinations of capitalists, whether engaged in manufacturing wool, cotton, or iron, would draw off labor from the cultivation of the soil, and cause large bodies of the producers to become consumers; and that roads and canals, connecting the West with the East, were effectual means of bringing the agricultural and manufacturing classes into closer proximity, to the serious limitation of the foreign commerce of the country, the checking of the growth of the navy, and the manifest injury of the planters.

This tariff and free trade controversy was

far from what it is now imagined to have been. People, on both sides, were often in great straits to know how to obtain a livelihood, much less to amass fortunes. The word *ruin* was no unmeaning phrase to the people of that day. The news, now, that a bank has failed, carries with it, to the depositors and holders of its notes, no stronger feelings of consternation, than did the report of the passage or repeal of tariff laws, then affect the minds of the opposing parties. We have spoken of the peculiar condition of the South in this respect. In the West, for many years, the farmers often received no more than *twenty-five cents*, and rarely over *forty cents* per bushel for their wheat, after conveying it, on horseback, or in wagons, not unfrequently a distance of fifty miles, to find a market. Other products were proportionally low in price; and such was the difficulty in obtaining money, that people could not pay their taxes but with the greatest sacrifices. So deeply were the people interested in these questions of national policy, that they became the basis of political action during several Presidential elections. This led to much vacillation in legislation on the subject, and gave alternately, to one and then to the other section of the Union, the benefits of its favorite policy.

The vote of the West, during this struggle, was of the first importance, as it possessed the balance of power, and could turn the scale at will. It was not left without inducements to co-operate with the South, in its measures for extending slavery, that it might create a market among the planters for its products. This appears from the particular efforts made by the Southern members of Congress, during the debate of 1824, to win over the West to the doctrines of Free trade.

Mr. McDuffie, of South Carolina, said: "I admit that the Western people are *embarrassed,* but I deny that they are *distressed,* in any other sense of the word. * * I am well assured that the permanent prosperity of the West depends more upon the improvement of the means of transporting their produce to market, and of receiving the returns, than upon every other subject to which the legislation of this government can be directed. * * Gentlemen (from the West) are aware that a very profitable trade is carried on by their constituents with the Southern country, in *live stock* of all descriptions, which they drive over the mountains and sell for cash. This extensive trade, which, from its peculiar character, more easily overcomes the difficulties of transportation than any that can be substituted in its place, is about to be put in jeopardy for the conjectural benefits of this measure. When I say this trade is about to be put in jeopardy, I do not speak unadvisedly. I am perfectly convinced that, if this bill passes, it will have the effect of inducing the people of the South, partly from the feeling and partly from the necessity growing out of it, to raise within themselves, the live stock which they now purchase from the West. * * If we cease to take the manufactures of Great Britain, she will assuredly cease to take our cotton to the same extent. It is a settled principle of her policy—a principle not only wise, but essential to her existence—to purchase from those nations that receive her manufactures, in preference to those who do not. You have, heretofore, been her best customers, and therefore, it has been her policy to purchase our cotton to the full extent of our demand for her manufactures. But, say gentlemen, Great Britain does not purchase your cotton from affection, but from interest. I grant it, sir; and that is the very reason of my decided hostility to a system which will make it her interest to purchase from other countries in preference to our own. It *is* her interest to purchase cotton, even at a higher price, from those countries which receive her manufactures in exchange. It is better for her to give a little more for cotton, than to obtain nothing for her manufactures. It will be remarked that the situation of Great Britain is, in this respect, widely different from that of the United States. The powers of her soil have been already pushed very nearly to the maximum of their productiveness. The productiveness of her manufactures, on the contrary, is as unlimited as the demand of the whole world. * * In fact, sir, the policy of Great Britain is not, as gentlemen seem to suppose, to secure the *home*, but the *foreign* market for her manufactures. The former she has without an effort. It is to attain the latter, that all her policy and enterprise are brought into requisition. The manufactures of that country are *the basis of her commerce;* our manufactures, on the contrary, are to be *the destruction of our commerce.* * * It can not be doubted, that, in pursuance of the policy of forcing her manufactures into foreign markets, she will, if deprived of a large portion of our custom, direct all her efforts to South America. That country abounds in a soil admirably adapted to the production of cotton, and will, for a century to come, import her manufactures from foreign countries."

Mr. HAMILTON, of South Carolina, said: "That the planters in his section shared in that depression which is common to every department of the industry of the Union, *excepting those from which we have heard the most clamor for relief.* This would be understood when it was known that sea-island cotton had fallen from 50 or 60 cents, to 25 cents—a fall even greater than that which has attended wheat, of which we had heard so much—as if the grain-growing section was the only agricultural interest which had suffered. * * While the planters of this region do not dread competition in the foreign markets on equal terms, from the superiority of their cotton, they entertain a well-founded apprehension, that the restrictions contemplated will lead to retaliatory duties on the part of Great Britain, which must end in ruin. * * In relation to our upland cottons, Great Britain may, without difficulty, in the course of a very short period, supply her wants from Brazil. * * How long the exclusive production, even of the sea-island cotton, will remain to our country, is yet a doubtful and interesting problem. The experiments that are making on the Delta of the Nile, if pushed to the Ocean, may result in the production of this beautiful staple, in an abundance which, in reference to other productions, has long blest and consecrated Egyptian fertility. * * We are told by the honorable Speaker (Mr. Clay), that our manufacturing establishments will, in a very short period, supply the place of the foreign demand. The futility, I will not say mockery of this hope, may be measured by one or two facts. First, the present consumption of cotton, by our manufactories, is about equal to one-sixth of our whole production. * * How long it will take to increase these manufactories to a scale equal to the consumption of this production, he could not venture to determine; but that it will be some years after the epitaph will have been written on the fortunes of the South, there can be little doubt." * * [After speaking of the tendency of increased manufactures in the East, to check emigration to the West, and thus to diminish the value of the public lands and prevent the growth of the Western States, Mr. H. proceeded thus:] "That portion of the Union could participate in no part of the bill, except in its burdens, in spite of the fallacious hopes that were cherished, in reference to cotton-bagging for Kentucky, and the woolen duty for Steubenville, Ohio. He feared that to the entire region of the West, no 'cordial drops of comfort' would come, even in the duty on foreign spirits. To a large portion of our people, who are in the habit of solacing themselves with Hollands, Antigua, and Cogniac, whisky, would still have 'a most villainous twang.' The cup, he feared, would be refused, though tendered by the hand of patriotism as well as conviviality. No, the West has but one interest, and that is, that its best customer, the South, should be prosperous."

Mr. RANKIN, of Mississippi, said: "With the West, it appears to me like a rebellion of the members against the body. It is true, we export, but the amount received from those exports is only apparently, largely in our favor, inasmuch as we are the consumers of your produce, dependent on you for our implements of husbandry, the means of sustaining life, and almost everything except our lands and negroes; all of which draws much from the apparent profits and advantages. In proportion as you diminish our exportations, you diminish our means of purchasing from you, and destroy your own market. You will compel us to use those advantages of soil and of climate which God and Nature have placed within our reach, and to live, as to you, as you desire us to live as to foreign nations—dependent on our own resources."

Mr. GARNETT, of Virginia, said: "The Western States can not manufacture. The want of capital (of which they, as well as the Southern States, have been drained by the policy of government), and other causes, render it impossible. The Southern States are destined to suffer more by this policy than any other — the Western next; but it will not benefit the aggregate population of any State. It is for the benefit of capitalists only. If persisted in, it will drive the South to ruin or resistance."

Mr. CUTHBERT, of Georgia, said: "He hoped the market for the cotton of the South was not about to be contracted within a little miserable sphere, the [home market], instead of being spread throughout the world. If they should drive the cotton-growers from the only source from whence their means were derived [the foreign market], they would be unable to take any longer their supplies from the West—they must contract their concerns within their own spheres, and begin to raise flesh and grain for their own consumption. The South was already under a severe pressure—if this measure went into effect, its distress would be consummated."

In 1828, the West found still very limited means of communication with the East. The opening of the New York canal, in 1825, created a means of traffic with the seaboard, to the people of the Lake region; but all of the remaining territory, west of the Alleghanies, had gained no advantages over those it had enjoyed in 1824, except so far as steamboat navigation had progressed on the Western rivers. In the debate preceding the passage of the tariff of 1828, usually termed the "Woolens' Bill," allusion is made to the condition of the West, from which we quote as follows:

Mr. Wickliffe, of Kentucky, said: "My constituents may be said to be a grain-growing people. They raise stock, and their surplus grain is converted into spirits. Where, I ask, is our market? * * Our market is where our sympathies should be, in the South. Our course of trade, for all heavy articles, is down the Mississippi. What breadstuffs we find a market for, are principally consumed in the States of Mississippi, Louisiana, South Alabama, and Florida. Indeed, I may say, these States are the consumers, at miserable and ruinous prices to the farmers in my State, of our exports of spirits, corn, flour, and cured provisions. * * We have had a trade of some value to the South in our stock. We still continue it under great disadvantages. It is a ready-money trade—I might say it is the only money trade in which we are engaged. * * Are gentlemen acquainted with the extent of that trade? It may be fairly stated at three millions per annum."

Mr. BENTON, urged the Western members to unite with the South, "for the purpose of enlarging the market, increasing the demand in the South and its ability to purchase the horses, mules, and provisions, which the West could sell nowhere else."

The tariff of 1828, created great dissatisfaction at the South. Examples of the expressions of public sentiment, on the subject, adopted at conventions, and on other occasions, might be multiplied indefinitely. Take a case or two, to illustrate the whole. At a public meeting in Georgia, held subsequently to the passage of the "Woolens' Bill," the following resolution was adopted:

Resolved, That to retaliate as far as possible upon our oppressors, our Legislature be requested to impose taxes, amounting to prohibition, on the hogs, horses, mules, and cotton-bagging, whisky, pork, beef, bacon, flax, and hemp cloth, of the Western, and on all the productions and manufactures of the Eastern and Northern States."

Mr. HAMILTON, of South Carolina, in a speech at the Waterborough Dinner, given subsequently to the passage of the tariff of 1828, said: "It becomes us to inquire what is to be our situation under this unexpected and disastrous conjuncture of circumstances, which, in its progress, will deprive us of the benefits of a free trade with the rest of the world, which formed one of the leading objects of the Union. Why, gentlemen, ruin, unmitigated ruin, must be our portion, if this system continues. * * From 1816 down to the present time, the South has been drugged, by the slow poison of the miserable empiricism of the prohibitory system, the fatal effects of which we could not so long have resisted, but for the stupendously valuable staples with which God has blessed us, and the agricultural skill and enterprise of our people."

The opening of the year 1832, found the parties to this controversy once more engaged in earnest debate, on the floor of Congress; and midsummer witnessed the passage of a new Tariff Bill, including the principle of Protection. Its enactment led to the movements in South Carolina toward *secession;* and, to avert the threatened evil, the Bill was modified, in the following year, so as to make it acceptable to the South; and, so as, also, to settle the policy of the Government for the succeeding nine years. A few extracts from the debates of 1832, will serve to show what were the sentiments of the members of Congress, as to the effects of the protective policy on the different sections of the Union, up to that date:

Mr. HAYNE, of South Carolina, said: "When the policy of '24 went into operation, the South was supplied from the West, through a single avenue, (the Saluda Mountain Gap,) with live stock, horses, cattle, and hogs, to the amount of considerably upward of a million of dollars a year. Under the pressure of the system, this trade has been regularly diminishing. It has already fallen more than one-half. * * * In consequence of the dire calamities which the system has inflicted on the South—blasting our commerce, and withering our prosperity—the West has been very nearly deprived of her best customer. * * And what was found to be the result of four years' experience at the South? Not a hope fulfilled; not one promise performed; and our condition infinitely worse than it had been four years before. Sir, the whole South rose up as one man, and protested against any farther experiment with this system. * * Sir, I seize the opportunity to dispel forever the delusion that the South can derive any compensation, in a home market, for the injurious operations of the protective system. * * What a spectacle do you even now exhibit to the world? A large portion of your fellow citizens, believing themselves to be grievously oppressed, by an unwise and unconstitutional system, are clamoring at your doors for justice; while another portion, supposing that they are enjoying rich bounties under it, are treating their complaints with scorn and contempt. * * This system may destroy the South, but it will not permanently advance the prosperity of the North. It may depress us, but can not elevate them. Beside, sir, if persevered in, it must annihilate that portion of the country from which the resources are to be drawn. And it may be well for gentlemen to reflect, whether adhering to this policy would not be acting like the man who 'killed the goose which laid the golden eggs.' Next to the Christian religion, I consider *Free Trade*, in its largest sense, as the greatest blessing that can be conferred on any people."

Mr. McDuffie, of South Carolina, said: "At the close of the late war with Great Britain, everything in the political and commercial changes resulting from the general peace, indicated unparalleled prosperity to the Southern

States, and great embarrassment and distress to those of the North. The nations of the Continent had all directed their efforts to the business of manufacturing; and all Europe may be said to have converted their swords into machinery, creating unprecedented demands for cotton, the great staple of the Southern States. There is nothing in the history of commerce that can be compared with the increased demand for this staple, notwithstanding the restrictions by which this Government has limited that demand. As cotton, tobacco, and rice are produced only on a small portion of the globe, while all other agricultural staples are common to every region of the earth, this circumstance gave the planting States very great advantages. To cap the climax of the commercial advantages opened to the cotton planters, England, their great and most valued customer, received their cotton under a mere nominal duty. On the other hand, the prospects of the Northern States were as dismal as those of the Southern States were brilliant. They had lost the carrying trade of the world, which the wars of Europe had thrown into their hands. They had lost the demand and the high prices which our own war had created for their grain and other productions; and, soon afterward, they also lost the foreign market for their grain, owing, partly, to foreign corn laws, but still more to other causes. Such were the prospects, and such the well founded hope of the Southern States at the close of the late war, in which they bore so glorious a part in vindicating the freedom of trade. But where are now these cheering prospects and animating hopes? Blasted, sir—utterly blasted—by the consuming and withering course of a system of legislation which wages an exterminating war against the blessings of commerce and the bounties of a merciful Providence; and which, by an impious perversion of language, is called "Protection." * * * I will now add, sir, my deep and deliberate conviction, in the face of all the miserable cant and hypocrisy with which the world abounds on the subject, that any course of measures which shall hasten the abolition of Slavery, by destroying the value of

Slave labor, will bring upon the Southern States the greatest political calamity with which they can be afflicted; for I sincerely believe, that when the people of those States shall be compelled, by such means, to emancipate their Slaves, they will be but a few degrees above the condition of slaves themselves. Yes, sir, mark what I say: when the people of the South cease to be masters, by the tampering influence of this Government, direct or indirect, they will assuredly be slaves. It is the clear and distinct perception of the irresistible tendency of this protective system to precipitate us upon this great moral and political catastrophe, that has animated me to raise my warning voice, that my fellow citizens may foresee, and, foreseeing, avoid the destiny that would otherwise befall them. * * * And here, sir, it is as curious as it is melancholy and distressing, to see how striking is the analogy between the Colonial vassalage to which the manufacturing States have reduced the planting States, and that which formerly bound the Anglo-American Colonies to the

British Empire. * * * England said to her American Colonies, You shall not trade with the rest of the world for such manufactures *as are produced in the mother country*. The manufacturing States say to their Southern Colonies, You shall not trade with the rest of the world for such manufactures as *we produce*, under a penalty of forty per cent. upon the value of every cargo detected in this illicit commerce; which penalty, aforesaid, shall be levied, collected, and paid, out of the products of your industry, to nourish and sustain ours."

Mr. CLAY, in referring to the condition of the country at large, said: "I have now to perform the more pleasing task of exhibiting an imperfect sketch of the existing state of the unparalleled prosperity of the country. On a general survey, we behold cultivation extended; the arts flourishing; the face of the country improved; our people fully and profitably employed, and the public countenance exhibiting tranquillity, contentment, and happiness. And, if we descend into particulars, we have the agreeable contemplation of a people out of debt; land rising slowly in value, but in a secure and salutary degree; a ready, though not an extravagant market for all the surplus productions of our industry; innumerable flocks and herds browsing and gamboling on ten thousand hills and plains, covered with rich and verdant grasses; our cities expanded, and whole villages springing up, as it were, by enchantment; our exports and imports increased and increasing; our tonnage, foreign and coastwise, swelled and fully occupied; the rivers of our interior animated by the perpetual thunder and lightning of countless steam-boats; the currency sound and abundant; the public debt of two wars nearly redeemed; and, to crown all, the public treasury overflowing, embarrassing Congress, not to find subjects of taxation, but to select the objects which shall be liberated from the impost. If the term of seven years were to be selected, of the greatest prosperity which this people have enjoyed since the establishment of their present Constitution, it would be exactly that period of seven years which immediately followed the passage of the tariff of 1824.

"This transformation of the condition of the country from gloom and distress to brightness and prosperity, has been mainly the work of American legislation, fostering American industry, instead of allowing it to be controlled by foreign legislation, cherishing foreign industry. The foes of the American system, in 1824, with great boldness and confidence, predicted, first, the ruin of the public revenue, and the creation of a necessity to resort to direct taxation. The gentleman from South Carolina, (General Hayne,) I believe, thought that the tariff of 1824 would operate a reduction of revenue to the large amount of eight millions of dollars; secondly, the destruction of our navigation; thirdly, the desolation of commercial cities; and, fourthly, the augmentation of the price of articles of consumption, and further decline in that of the articles of our exports. Every prediction which they made has failed—utterly failed. * * It is now proposed to abolish the system to which we owe so much of the public prosperity. * * Why, sir, there is scarcely an interest—scarcely a vocation in society—which is not embraced by the beneficence of this system. * * The error of the opposite argument, is in assuming one thing, which, being denied, the whole fails; that is, it assumes that the *whole* labor of the United States would be profitably employed without manufactures. Now, the truth is, that the system *excites* and *creates* labor, and this labor creates wealth, and this new wealth communicates additional ability to consume; which acts on all the objects contributing to human comfort and enjoyment. * * I could extend and dwell on the long list of articles—the hemp, iron, lead, coal, and other items — for which a demand is created in the home market by the operation of the American system; but I should exhaust the patience of the Senate. *Where, where* should we find a market for all these articles, if it did not exist at home? What would be the condition of the largest portion of our people, and of the territory, if this home market were annihilated? How could they be supplied with objects of prime necessity? What would not be the certain and inevitable decline in the price of all these articles, but for the home market?"

But we must not burden our pages with further extracts. What has been quoted affords the principal arguments of the opposing parties, on the points in which we are interested, down to 1832. The adjustment, in 1833, of the subject until 1842, and its subsequent agitation, are too familiar, or of too easy access to the general reader, to require a notice from us here.

The results of the contest, in relation to Protection and Free Trade, have been more or less favorable to all parties. This has been an effect, in part, of the changeable character of our legislation; and, in part, of the occurrence of events over which politicians had no control. The manufacturing States, while protection lasted, succeeded in placing their establishments upon a comparatively permanent basis; and, by engaging largely in the manufacture of cottons, as well as woolens, have rendered home manufactures, practically, very advantageous to the South. Our cotton factories, in 1850, consumed as much cotton as those of Great Britain did in 1831; thus affording indications, that, by proper encouragement, they may be multiplied so as to consume the whole crop of the country. The cotton and woolen factories, in 1850, employed over 130,000 work hands, and had $102,619,581 of capital invested in them. They thus afford an important market to the farmer, and, at the same time, have become an equally important auxiliary to the planter. They may yet afford him the only market for his cotton.

The cotton planting States, toward the close of the contest, found themselves rapidly accumulating strength, and approximating the accomplishment of the grand object at which they aimed—the monopoly of the cotton markets of the world. This success was due, not so much to any triumph over the North—to any prostration of our manufacturing interests—as to the general policy of other nations. All rivalry to the American planters, in the West Indies, was removed by emancipation; as, under freedom, the cultivation of cotton was nearly abandoned. Mehemet Ali had become imbecile, and the indolent Egyptians neglected its culture. The South Americans, after achieving their independence, were more readily enlisted in military forays, than in the art of agriculture, and they produced little cotton for export. Brazil and India both supplied to Europe considerably less in 1838 than they had done in 1820; and the latter country made no material increase afterward, except when her chief customer, China, was at war, or prices were above the average rates in Europe. While the cultivation of cotton was thus stationary or retrograding, everywhere outside of the United States, England and the Continent were rapidly increasing their consumption of the article, which they nearly doubled from 1835 to 1845; so that the demand for the raw material called loudly for its increased production. Our planters gathered a rich harvest of profits by these events.

But this is not all that is worthy of note, in

this strange chapter of providences. No prominent event occurred, but conspired to advance the prosperity of the cotton trade, and the value of American Slavery. Even the very depression suffered by the manufacturers and cultivators of cotton, from 1825 to 1829, served to place the manufacturing interests upon the broad and firm basis they now occupy. It forced the Planters into the production of their cotton at reduced rates; and led the Manufacturers to improve their machinery, and reduce the price of their fabrics low enough to sweep away all *Household manufacturing*, and secure to themselves the monopoly of clothing the civilized world. This was the object at which the British manufacturers had aimed, and in which they had been eminently successful. The growing manufactures of the United States, and of the Continent of Europe, had not yet sensibly affected their operations.

There is still another point requiring a passing notice, as it may serve to explain some portions of the history of Slavery, not so well understood. It was not until events diminishing the foreign growth of cotton, and enlarging the demand for its fabrics, had been extensively developed, that the older cotton-growing States, became willing to allow Slavery extension in the Southwest; and, even then, their assent was reluctantly given — the markets for cotton, doubtless, being considered sufficiently limited for the territory under cultivation. Up to 1824 the Indians held over thirty-two millions of acres of land in Georgia, Mississippi, and Alabama, and over twenty millions of acres in Florida, Missouri, and Arkansas; which was mostly retained by them as late as 1836. Although the States interested had repeatedly urged the matter upon Congress, and some of them even resorted to forcible means to gain possession of these Indian lands, the Government did not fulfill its promise to remove the Indians until 1836.

The older States, however, had found, by this time, that the foreign and home demand for cotton was so rapidly increasing, that there was little danger of over-production; and that they had, in fact, secured to themselves the monopoly of the foreign markets. Beside this, the Abolition movement, at that moment, had assumed its most threatening aspect, and was demanding the destruction of Slavery or the dissolution of the Union. Here was a double motive operating to produce harmony in the ranks of Southern politicians, and to awaken the fears of many, North and South, for the safety of the Government. Here, also, was the origin of the determination, in the South, to extend Slavery, by the annexation of territory, so as to gain the political preponderance in the National Councils, and protect its interests against the interference of the North.

It was not the increased demand for cotton, alone, that served as a protection to the older States. The extension of its cultivation, in the degree demanded by the wants of commerce, could only be affected by a corresponding increased supply of Provisions. Without this it could not increase, except by enhancing their price to the injury of the older States. This food did not fail to be in readiness, so soon as it was needed. Indeed, much of it had long been awaiting an outlet to a profitable market. Its surplus, too, had been materially increased, by the Temperance movement in the North, which had checked, somewhat, the distillation of grain.

The West, which had long looked to the East for a market, had its attention now turned to the South, as the most certain and convenient mart for the sale of its products — the Planters affording to the Farmers, the markets they had in vain sought from the Manufacturers. In the meantime steamboat navigation was acquiring perfection on the Western rivers — the great natural outlets for Western products — and became a means of communication between the Northwest and the Southwest, as well as with the trade and commerce of the Atlantic cities. This gave an impulse to industry and enterprise, west of the Alleghanies, unparalleled in the history of the country. While, then, the bounds of Slave labor were extending from Virginia, the Carolinas, and Georgia, westward, over Tennessee, Alabama, Mississippi, and Arkansas, the area of Free labor was enlarging, with equal rapidity, in the Northwest, throughout Ohio, Indiana, Illinois, and Michigan. Thus within these Provision and Cotton regions, were the forests cleared away, or the prairies broken up, simultaneously, by these old antagonistic forces, opponents no longer, but harmonized by the fusion of their interests — the connecting link between them being the steamboat. Thus, also, was a *tripartite alliance* formed, by which the Western Farmer, the Southern Planter, and the English Manufacturer, became united in a common bond of interest — the whole giving their support to the doctrines of Free Trade.

This active commerce between the West and South, however, soon caused a rivalry in the East, that pushed forward improvements, by States or Corporations, to gain a share in the Western trade. These improvements, as completed, gave to the West a choice of markets, so that its Farmers could elect whether to feed the slave who grows the cotton, or the operatives who are engaged in its manufacture. But this rivalry did more. The competition for Western products enhanced their price, and stimulated their more extended cultivation. This required an enlargement of the markets; and the extension of Slavery became essential to Western prosperity.

We have not reached the end of the alliance between the Western Farmer and Southern Planter. The emigration which has been filling Iowa and Minnesota, and is now rolling like a flood into Kansas and Nebraska, is but a repetition of what has occurred in the other Western States and Territories. Agricultural pursuits are highly remunerative, and tens of thousands of men of moderate means, or of no means, are cheered along to where none forbids them land to till. For the last few years, public improvements have called for vastly more than the usual share of labor, and augmented the consumption of Provisions. The foreign demand added to this, has increased their price beyond what the Planter can afford to pay. For many years Free and Slave labor maintained an even race in their Western progress. Of late the Freemen have begun to lag behind, while Slavery has advanced by several degrees of longitude. Free labor must be made to keep pace with it. There is an urgent necessity for this. The demand for cotton is increasing in a ratio greater than can be supplied by the American planters, unless by a corresponding increased production. They must meet this increasing demand, or its cultivation will be facilitated elsewhere, and our monopoly of the European markets be interrupted. This can only be effected by concentrating the greatest possible number of the slaves upon the cotton plantations. Hence they must be supplied with provisions.

This is the present aspect of the Provision question, as it regards Slavery extension. Prices are approximating the maximum point, beyond which our provisions can not be fed to slaves. Such a result was not anticipated by Southern statesmen, when they had succeeded in overthrowing the Protective policy, destroying the United States Bank, and establishing the Sub-Treasury system. And why has this occurred? The mines of California prevented both the Free-Trade Tariff,* and the Sub-Treasury scheme

* The Tariff of 1846, under which our imports are now made, approximates the Free Trade principles very closely.

from exhausting the country of the precious metals, extinguishing the circulation of Bank Notes, and reducing the prices of agricultural products to the specie value. At the date of the passage of the Nebraska Bill, the multiplication of provisions, by their more extended cultivation, was the only measure left that could produce a reduction of prices, and meet the wants of the planters. The Canadian reciprocity treaty, since secured, will bring the products of the British North American Colonies, free of duty, into competition with those of the United States, when prices, with us, rule high, and tend to diminish their cost; but in the event of scarcity in Europe, or of foreign wars, the opposite results may occur, as our products, in such times, will pass, free of duty, through these Colonies, into the foreign market. It is apparent, then, that nothing short of extended Free-labor cultivation, far distant from the seaboard, where the products will bear transportation to none but Southern markets, can fully secure the Cotton interests from the contingencies that so often threaten them with ruinous embarrassments.

In fact, such a depression of our cotton interests has only been averted by the advanced prices which cotton has commanded, for the last few years, in consequence of the increased European demand, and its diminished cultivation abroad.

The dullest intellect can not fail, now, to perceive the rationale of the Kansas-Nebraska movement. The political influence which these Territories will give to the South, if secured, will be of the first importance to perfect its arrangements for future Slavery extension—whether by divisions of the larger States and Territories now secured to the institution, its extension into territory hitherto considered free, or the acquisition of new territory to be devoted to the system, so as to preserve the balance of power in Congress. When this is done, Kansas and Nebraska, like Kentucky and Missouri, will be of little consequence to slaveholders, compared with the cheap and constant supply of provisions they can yield. Nothing, therefore, will so exactly coincide with Southern interests, as a rapid emigration of freemen into these new Territories. Free labor, doubly productive over

Slave labor, in grain-growing, must be multiplied within their limits, that the cost of provisions may be reduced, and the extension of Slavery and the growth of cotton suffer no interruption. The present efforts to plant them with Slavery, are indispensable to produce sufficient excitement to fill them speedily with a free population; and if this whole movement has been a Southern scheme to cheapen provisions, and increase the ratio of the production of sugar and cotton, as it most unquestionably will do, it surpasses the statesman-like strategy which forced the people into an acquiescence in the annexation of Texas.

And should the Anti-Slavery voters succeed in gaining the political ascendency in these Territories, and bring them as free States triumphantly into the Union; what can they do, but turn in, as all the rest of the Western States have done, and help to feed slaves, or those who manufacture or who sell the products of the labor of slaves. There is no other resource left, either to them or ourselves, without an entire change in almost every branch of business and of domestic economy. Look at your bills of dry-goods for the year, and what do they contain? At least three-fourths of the amount are French, English, or American cotton fabrics, woven from Slave-labor cotton. Look at your bills for groceries, and what do they contain? Coffee, sugar, molasses, rice—from Brazil, Cuba, Louisiana, Carolina; while only a mere fraction of them are from Free-labor countries. As now employed, our dry-goods merchants and grocers constitute an immense army of agents for the sale of fabrics and products, coming directly or indirectly, from the hand of the slave; and all the remaining portion of the people, free colored, as well as white, are exerting themselves, according to their various capacities, to gain the means of purchasing the greatest possible amount of these commodities. Nor can the country, at present, by any possibility, pay for the amount of foreign goods consumed, but by the labor of the slaves of the planting States. This can not be doubted for a moment. Here is the proof:

Commerce supplied us, in 1853, with foreign articles, for consumption, to the value of

$250,420,187, and accepted in exchange, of our provisions, to the value of but $33,809,126; while the products of our Slave labor, manufactured and unmanufactured, paid to the amount of $133,648,603, on the balance of this foreign debt. This, then, is the measure of the ability of the farmers and planters, respectively, to meet the payment of the necessaries and comforts of life, supplied to the country by its foreign commerce. The farmer pays, or seems only to pay, $33,800,000, while the planter has a broad credit, on the account, of $133,600,000.

But is this seeming productiveness of Slavery real, or is it only imaginary? Has the system such capacities, over the other industrial interests of the nation, in the creation of wealth, as these figures indicate? Or, are these results due to its intermediate position between the agriculture of the country and its foreign commerce? These are questions worthy of consideration. Were the planters left to grow their own provisions, they would, as already intimated, be unable to produce any cotton for export. That their present ability to export so extensively, is in consequence of the aid they receive from the North, is proved by facts such as these:

In 1820, the cotton-gin had been a quarter of a century in operation, and the culture of cotton was then as well understood as at present. The North, though furnishing the South with some live stock, had scarcely begun to supply it with Provisions, and the planters had to grow the food, and manufacture much of the clothing for their slaves. In that year the cotton crop equaled 109 lbs. to each slave in the Union, of which 83 lbs. per slave were exported. In 1830 the exports of the article had risen to 143 lbs., in 1840 to 295 lbs., and in 1853 to 337 lbs. per slave. The total cotton crop of 1853, equaled 485 lbs. per slave—making both the production and export of that staple, in 1853, more than four times as large, in proportion to the Slave population, as they were in 1820.* Had the

* The progressive increase is indicated by the following figures:

	1820	1830	1840	1853
Total slaves in U. S.,	1,538,098	2,009,043	2,487,356	3,296,408
Cot. Exp'd, lbs.,	127,800,000	298,459,102	743,941,061	1,111,570,370
Av'ge ex. to each slave, lbs.,	83	143	295	337

planters, in 1853, been able to produce no more cotton, per slave, than in 1820, they would have grown but 359,308,472 lbs., instead of the actual crop of 1,600,000,000 lbs.; and would not only have failed to supply any for export, but have fallen short of the home demand, by nearly 130,000,000 lbs., and been *minus* the total crop of that year, by 1,240,690,000 lbs.

In this estimate, some allowance, perhaps, should be made, for the greater fertility of the new lands, more recently brought under cultivation; but the difference, on this account, can not be equal to the difference in the crops of the several periods, as the lands, in the older States, in 1820, were yet comparatively fresh and productive.

Again, the dependence of the South upon the North, for its provisions, may be inferred from such additional facts as these: The "Abstract of the Census," for 1850, shows, that the production of wheat, in Florida, Alabama, Mississippi, Louisiana, Arkansas, and Texas, averaged, the year preceding, very little more than a peck, (it was 27-100 of a bushel,) to each person within their limits. These States must purchase flour largely, but to what amount we can not determine. The shipments of provisions from Cincinnati to New Orleans and other down river ports, show that large supplies leave that city for the South ; but what proportion of them is taken for consumption by the planters, must be left, at present, to conjecture. These shipments, as to a few of the prominent articles, for the four years ending August 31, 1854, averaged, annually, the following amounts:

Wheat flour	bbls.	385,204
Pork and bacon, in bulk and in barrels*	lbs.	21,095,930
" " hogsheads	hhds.	20,767
" " tierces,	tcs.	15,478
Whisky†	gals.	8,115,360

Cincinnati also exports eastward, by canal, river, and railroad, large amounts of these productions. The towns and cities westward send more of their products to the South, as their distance increases the cost of transportation to the East. But, in the absence of full statis-

* Barrels estimated to contain 196 lbs. each.

† Barrels estimated at 40 gallons each.

tics, it is not necessary to make additional statements.

From this view of the subject, it appears that Slavery is not a self-sustaining system, independently remunerative; but that it attains its importance to the nation and to the world, by standing as an agency, intermediate between the grain-growing States and foreign commerce. As the distillers of the West transformed the surplus grain into whisky, that it might bear transport, so Slavery takes the products of the North, and metamorphoses them into cotton, that they may bear export.

It seems, indeed, when the whole of the facts brought to view are considered, that American Slavery, though of little force unaided, yet, properly sustained, is the great central power, or energizing influence, not only of nearly all the industrial interests of our own country, but also of those of Great Britain and much of the Continent; and that, if stricken from existence, the whole of these interests, with the advancing civilization of the age, would receive a shock that must retard their progress for years to come.

This is no exaggerated picture of the present imposing power of Slavery. It is literally true. Southern men believed that the Protective Tariff would have paralyzed it—would have destroyed it. But the Abolitionists, led off by politicians and editors, who advocated Free Trade, were made the instruments of its overthrow. No such extended mining and manufacturing, as the Protective system would have created, has now any existence in the Union. Under it, more than one hundred and sixty millions in value, of the foreign imports for 1853, would have been produced in our own country. But Free Trade is dominant: the South has triumphed in its warfare with the North: the political power passed into its hands with the defeat of the Father of the Protective Tariff, ten years since, in the last effort of his friends to elevate him to the Presidency: the Slaveholding and commercial interests then gained the ascendency, and secured the power of annexing territory at will; the nation has become rich in commerce, and unbounded in ambition for territorial aggrandizement: the people acquiesce in the measures of Government, and are proud of its influence in the world; nay, more, the peaceful aspect of the nations has been changed, and the policy of our own country must be modified to meet the exigencies that may arise.

One word more on the point we have been considering. With the defeat of Mr. CLAY, came the immediate annexation of Texas, and, as he predicted, the war with Mexico. The results of these events let loose from its attachments a mighty avalanche of emigration and of enterprise, under the rule of the Free Trade policy, then adopted, which, by the golden treasures it yields, renders that system, thus far, self-sustaining, and able to move on, as its friends believe, with a momentum that forbids any attempt to return again to the system of Protection. Whether the Tariff controversy is permanently settled, or not, is a question about which we shall not speculate. It may be remarked, however, that one of the leading parties in the North, gave in its adhesion to Free Trade many years since, and still continues to vote with the South. The leading Abolition paper, too, ever since its origin, has advocated the Southern Free Trade system; and thus, in defending the cause it has espoused, as was said of a certain General in the Mexican war, its editor has been digging his ditches on the wrong side of his breastworks. To say the least, his position is a very strange one, for a man who professes to labor for the overthrow of American Slavery. It would be as rational to pour oil upon a burning edifice, to extinguish the fire, as to attempt to overthrow that system under the rule of Free Trade.

All these things together have paralyzed the advocates of the protection of Free labor, at present, as fully as the North has thereby been shorn of its power to control the question of Slavery. Indeed, from what has been said of the present position of American Slavery, in its relations to the other industrial interests of the country, and of the world, there is no longer any doubt that it now completes the *home market*, so zealously urged as essential to the prosperity of the Agricultural population of the country: and which, it was supposed, could only be created by the multiplication of domestic manufactures. This desideratum being gained, the great majority of the people have nothing more to ask, but seem desirous that our foreign commerce shall be cherished; that the cultivation of cotton and sugar shall be extended; that the nation shall become cumulative as well as progressive; that as Despotism is striving to spread its raven wing over the earth, Freedom must strengthen itself for the protection of the liberties of the world; that while three millions of Africans, only, are held to involuntary servitude for a time, to sustain the system of Free Trade, the freedom of hundreds of millions is involved in the preservation of the American Constitution; and that as African emancipation, in every experiment made, has thrown a dead weight upon Anglo-Saxon progress, the colored people must wait a little, until the general battle for the liberties of the civilized nations is gained, before the universal elevation of the barbarous tribes can be achieved. This work, it is true, has been commenced at various outposts in heathendom, by the missionary, but is impeded by numberless hinderances; and these obstacles to the progress of Christian civilization, doubtless will continue, until the friends of civil and religious liberty shall triumph in nominally Christian countries, and, with the wealth of the nations at command, instead of applying it to purposes of war, shall devote it to sweeping away the darkness of superstition and barbarism from the earth, by extending the knowledge of Science and Revelation to all the families of man.

But we must hasten.

There are none who will deny the truth of what is said of the present strength and influence of Slavery, however much they may have deprecated its acquisition of power. There are none who think it practicable to assail it, successfully, by political action, in the States where it is already established by law. The struggle against the system, therefore, is narrowed down to an effort to prevent its extension into territory now free; and this contest is limited to the people of the Territories themselves. The question is thus taken out of the hands of the people at large, and they are cut off from all control of Slavery, both in the States and Territories. Hence it is, that the American people are considering the propriety of banishing this distracting question from national politics, and demanding of their statesmen that there shall no longer be any delay in the adoption of measures to sustain the Constitution and laws of our glorious Union, against all its enemies, whether domestic or foreign.

The policy of adopting this course, may be liable to objection; but it does not appear to arise from any disposition to prove recreant to the cause of philanthropy, that the people of the Free States are resolving to divorce the Slavery question from all connection with political movements. It is because they now find themselves wholly powerless, as did the Colonizationists, forty years since, in regard to emancipation, and are thus forced into a position of neutrality upon that subject. A word on this point. The friends of Colonization, in the outset of that enterprise, found themselves shut up to the necessity of creating a Republic on the shores of Africa, as the only hope for the Free colored people—the further emancipation of the slaves, by State action, having become impracticable. After nearly forty years of experimenting with the free colored people, by others, Colonizationists still find themselves circumscribed in their operations, to their original design of building up the Republic of Liberia, as the only rational hope of the elevation of the African race—the prospects of general emancipation being a thousand-fold more gloomy in 1855 than they were in 1817. But to return. The people at large, too, begin not only to realize their own want of power over the institution of Slavery, and the futility of any measures hitherto adopted to arrest its progress, and elevate the free colored people; but they have also discovered agencies at work, heretofore overlooked, except by few, which are tending to sap the foundations of our Free Institutions, and to subject us to influences that have crushed the liberties of Europe, and which, if permitted to become dominant here, will blot out our happy Republic, and, with it, the liberties of the world.

In such a crisis as this, shall the friends of the Union be rebuked, if they determine to take a position of neutrality, in politics, on the subject of Slavery; while, at the same time, they offer to guarantee the Free colored people a Republic of their own, where they may equal other races, and aid in redeeming a continent from the woes it has suffered for thousands of years!

3. The social and moral condition of the free people of color, in the British colonies, and in the United States; and the new field opening in Liberia for the display of their powers.

We have noticed the social and moral condition of the free colored people, from the days of FRANKLIN, to the projection of Colonization. We have also glanced at the main facts in relation to the Abolition warfare upon Colonization, and its success in paralyzing the enterprise. This demands a more extended notice. The most serious injury from this hostility, sustained by the cause of Colonization, was the prejudice created, in the minds of the more intelligent free colored men, against emigration to Liberia. The Colonization Society had expressed its belief in the *natural equality* of the blacks and whites; and that there were a sufficient number of educated, upright, free colored men, in the United States, to establish and sustain a Republic on the coast of Africa, "whose citizens, rising rapidly in the scale of existence, under the stimulants to noble effort by which they would be surrounded, might soon become equal to the people of Europe, or of European origin—so long their masters and oppressors." These were the sentiments of the first Report of the Colonization Society, and often repeated since. Its appeals were made to the moral and intelligent of the Free colored people; and, with their co-operation, the success of its scheme was considered certain. But the very persons needed to lead the enterprise, were, mostly, persuaded to reject the proffered aid, and the Society was left to prosecute its plans with such materials as offered. In consequence of this opposition, it was greatly embarrassed, and made less progress in its work of African redemption, than it must have done under other circumstances. Had three-fourths of its emigrants been the enlightened, free colored men of the country, a dozen Liberias might now gird the coast of Africa, where but one exists; and the Slave trader be entirely excluded from its shores. Doubtless, a wise Providence has governed here, as in other human affairs, and may have permitted this result, to show how speedily even semi-civilized men can be elevated under American Protestant Free Institutions. The great body of emigrants to Liberia, and nearly all the leading men who have sprung up in the Colony, and contributed most to the formation of the Republic, went out from the very midst of Slavery; and yet, what encouraging results! It has been a sad mistake to oppose American Colonization, and thus to retard Africa's redemption!

But how has it fared with the Free colored people elsewhere? The answer to this question will be the solution of the inquiry, What has Abolitionism accomplished by its hostility to

Colonization, and what is the condition of the free colored people, whose interests it volunteered to promote, and whose destinies it attempted to control?

The Abolitionists themselves shall answer this question. The colored people shall see what kind of commendations their tutors give them, and what the world is to think of them, on the testimony of their particular friends.

The concentration of a colored population in Canada, is the work of American Abolitionists. In 1848, Rev. E. SMITH, a prominent Abolitionist of Ohio, was acting as their agent, in collecting funds for their relief. In an appeal for aid, published in the *Clarion of Freedom*, Feb. 18, he represented them as "destitute of education, and, like all other uneducated persons, having no great appreciation of its value, and not making the exertions they should to secure it to themselves or their children."

The American Missionary Association, is the organ of the Abolitionists, for the spread of a Gospel untainted, it is claimed, by contact with Slavery. Out of four stations under its care, in Canada, at the opening of 1853, but one school, that of MISS LYON, remained at its close. All the others were abandoned, and all the missionaries had asked to be released,[*] as we are informed by its Seventh Annual Report, mainly, for the reasons stated in the following extract, page 49:

"The number of missionaries and teachers in Canada, with which the year commenced, has been greatly reduced. Early in the year, Mr. KIRKLAND wrote to the Committee, that the opposition to white missionaries, manifested by the colored people of Canada, had so greatly increased, by the interested misrepresentations of ignorant colored men, pretending to be ministers of the Gospel, that he thought his own and his wife's labors, and the funds of the Association, could be better employed elsewhere."

It is not our purpose to multiply testimony on this subject, but simply to afford an index to the condition of the colored people, as described

[*] Mr. WILSON, the Missionary at St. Catherines, still remained there, but not under the care of the Association.

by Abolition pens, best known to the public. West India Emancipation, under the guidance of English Abolitionists, has always been viewed as the grand experiment, which was to convince the world of the capacity of the colored man to rise, side by side, with the white man. We shall let the friends of the system testify as to its results. Opening, again, the Seventh Annual Report of the American Missionary Association, page 30, we find it written:

"One of our missionaries, in giving a description of the moral condition of the people of Jamaica, after speaking of the licentiousness which they received as a legacy from those who denied them the pure joys of holy wedlock, and trampled upon and scourged chastity, as if it were a fiend to be driven out from among men — that enduring legacy which, with its foul, pestilential influence, still blights, like the mildew of death, everything in society that should be lovely, virtuous, and of good report; and alluding to their intemperance, in which they have followed the example set by the Governor in his palace, the Bishop in his robes, statesmen and judges, lawyers and doctors, planters and overseers, and even professedly Christian ministers; and the deceit and falsehood which oppression and wrong always engender, says: 'It must not be forgotten that we are following in the wake of the accursed system of *Slavery*— a system that *unmakes man*, by warring upon his conscience and crushing his spirit, leaving naught but the shattered wrecks of humanity behind it. If we may but gather up some of these floating fragments, from which the image of God is well nigh effaced, and pilot them safely to that better land, we shall not have labored in vain. But we may *hope to do more.* The chief fruit of our labors is to be sought in the *future* rather than in the *present.*' It should be remembered, too, (continues the Report,) that there is but a small part of the population yet brought within the reach of the influence of enlightened Christian teachers, while the great mass by whom they are surrounded are but little removed from actual heathenism." Another missionary, page 33, says, it is the opinion of all intelligent Christian men, that "nothing save the furnishing of the people with ample means of education and religious instruction will save them from relapsing into a state of barbarism." And another, page 36, in speaking of certain cases of discipline, for the highest form of crime, under the seventh commandment, says: "There is *nothing* in public sentiment to save the youth of Jamaica in this respect."

The missions of this Association, in Jamaica, differ scarcely a shade from those among the actual heathen. On this point, the Report, near its close, says:

"For most of the adult population of Jamaica, the unhappy victims of long years of oppression and degradation, our missionaries have great fear. Yet for even these there may be hope, even though with trembling. But it is around the youth of the island that their brightest hopes and anticipations cluster; from them they expect to gather their principal sheaves for the great Lord of the harvest."

Thus far we have drawn upon the *American Missionary Association.* Next we turn to the *Annual Report of the American and Foreign*

Anti-Slavery Society, 1853, which discourses thus, in its own language, and in quotations which it indorses: *

"The friends of emancipation in the United States have been disappointed in some respects at the results in the West Indies, because they expected too much. A nation of slaves can not at once be converted into a nation of intelligent, industrious, and moral freemen." * * "It is not too much, even now, to say of the people of Jamaica, * * their condition is exceedingly degraded, their morals woefully corrupt. But this must by no means be understood to be of universal application. With respect to those who have been brought under a healthful educational and religious influence, *it is not true.* But as respects the great mass, whose humanity has been ground out of them by cruel oppression — whom no good Samaritan hand has yet reached — how could it be otherwise? We wish to turn the tables; to supplant oppression by righteousness, insult by compassion and brotherly-

---

* Page 170.

kindness, hatred and contempt by love and winning meekness, till we allure these wretched ones to the hope and enjoyment of manhood and virtue."* * * "The means of education and religious instruction are better enjoyed, although but little appreciated and improved by the great mass of the people. It is also true, that the moral sense of the people is becoming somewhat enlightened. * * But while this is true, yet their moral condition is very far from being what it ought to be. * * It is exceeding dark and distressing. Licentiousness prevails to a most alarming extent among the people. * * The almost universal prevalence of intemperance is another prolific source of the moral darkness and degradation of the people. The great mass, among all classes of the inhabitants, from the Governor in his palace to the peasant in his hut—from the bishop in his gown to the beggar in his rags — are all slaves to their cups."†

* Extract from the report of a missionary, quoted in the Report, page 172.

† Extract from the report of another missionary, page 171, of the Report.

This is the language of American Abolitionists, going out under the sanction of their Annual Reports. Lest it may be considered as too highly colored, we add the following from the London Times, of near the same date. In speaking of the results of emancipation, in Jamaica, it says:

"The negro has not acquired with his freedom any habits of industry or morality. His independence is but little better than that of an uncaptured brute. Having accepted few of the restraints of civilization, he is amenable to few of its necessities; and the wants of his nature are so easily satisfied, that at the current rate of wages, he is called upon for nothing but fitful or desultory exertion. The blacks, therefore, instead of becoming intelligent husbandmen, have become vagrants and squatters, and it is now apprehended that with the failure of cultivation in the island will come the failure of its resources for instructing or controlling its population. So imminent does this consummation appear, that memorials have been signed by classes of colonial society hitherto standing aloof from politics, and not only the bench and the bar, but the bishop, clergy, and ministers of all denominations in the island, without exceptions, have recorded their conviction, that, in the absence of timely relief, the religious and educational institutions of the island must be abandoned, and the masses of the population retrograde to barbarism."

One of the editors of the *New York Evening Post*, Mr. BIGELOW, a few years since, spent a winter in Jamaica, and continues to watch, with anxious solicitude, as an Anti-Slavery man, the developments taking place among its emancipated freedmen. In reviewing the returns published by the Jamaica House of Assembly, in 1853, in reference to the ruinous decline in the Agriculture of the Island, and stating the enormous quantity of lands thrown out of cultivation, since 1848, the *Post* says:

"This decline has been going on from year to year, daily becoming more alarming, until at length the Island has reached what would appear to be the last profound of distress and misery, * * when thousands of people do not know, when they rise in the morning, whence or in what manner they are to procure bread for the day."

After such an array of testimony from Abolition authorities, we may venture to present some corroborative evidence from other sources. Governor Wood, of Ohio, on his way to Valparaiso, in 1853, thus describes what he witnessed, at Kingston, Jamaica, while the steamer remained in that port:

"We saw many plantations, the buildings dilapidated,—fields of sugar-cane half-worked and apparently poor, and nothing but that which will grow without the labor of man, appeared luxuriant and flourishing. The island itself is of great fertility, one of the best of the Antilles; but all the large estates upon it are now fast going to ruin. In the harbor were not a dozen ships of all nations—no business was doing, and everything you heard spoken was in the language of complaint. Since the blacks have been liberated they have become indolent, insolent, degraded, and dishonest. They are a rude, beastly set of vagabonds, lying naked about the streets, as filthy as the Hottentots, and, I believe, worse."

Bishop Kip, of the Protestant Episcopal Church, on his passage to California, in 1853, bears this testimony as to what he witnessed at the same port, while the steamer stopped to take in coal:

"The streets are crowded with the most wretched-looking negroes to be seen on the face of the earth. Lazy, shiftless and diseased, they will not work, since the manumission act has freed them. Even coaling the steamer is done by women. About a hundred march on board in a line with tubs on their heads (tubs and coal together weighing about 90 pounds), and with a wild song empty them into the hold. The men work a day, and then live on it a week. The depth of degradation to which the negro population has sunk, is, we are told, indescribable."

The foregoing testimony is conclusive, as to the results of emancipation in the West Indies. It fully confirms the opinions of Franklin, that freedom, to unenlightened slaves, must be accompanied with the means of intellectual and moral elevation, otherwise it may be productive of serious evils to themselves and to society. It also sustains the views entertained by Southern Slaveholders, that emancipation, unaccompanied by the colonization of the slaves, could be of no value to the blacks, while it would entail a ruinous burden upon the whites. These facts must not be overlooked in the projection of plans for emancipation, as none can receive the sanction of Southern men, which does not embrace in it the removal of the colored people. With the example of West India emancipation before them, and the results of which have been closely watched by them, it can not be expected that Southern statesmen will risk the liberation of their slaves, except on these conditions.

In turning to the condition of our own Free colored people, who rejected homes in Liberia, we approach a most important subject. They have been under the guardianship of their Abolition friends, ever since that period, and have cherished feelings of determined hostility to Colonization. What have they gained by this hostility? What has been accomplished for them by their Abolition friends, or what have they done for themselves? Those who took refuge in Liberia, have built up a Republic of their own, and are recognized as an independent nation, by five of the great governments of the earth. But what has been the progress of those who remained behind, in the vain hope of rising to an equality with the whites, and of assisting in abolishing American Slavery?

We offer no opinion, here, of our own, as to the present social and moral condition of the Free colored people in the North. What it was at the time of the founding of Liberia, has already been shown. On this subject we might quote largely from the proceedings of their conventions, and the writings of their editors, so as to produce a dark picture indeed; but this would be cruel, as their voices are but the wailings of noble, sensitive, and benevolent hearts, while weeping over the moral desolations that have overwhelmed their people. Nor shall we multiply testimony on the subject; but in this, as in the case of Canada and the West Indies, allow the Abolitionists to speak of their own schemes. One witness only, the most calm and reliable of them all, need be quoted. The Hon. GERRITT SMITH, in his letter to GOVERNOR HUNT, of New York, in 1852, while speaking of his ineffectual efforts, for fifteen years past, to prevail upon the Free colored people to betake themselves to mechanical and agricultural pursuits, says:

"Suppose, moreover, that during all these fifteen years, they had been quitting the cities, *where the mass of them rot both physically and morally*, and had gone into the country to become farmers and mechanics—suppose, I say, all this — and who would have the hardihood to affirm that the Colonization Society lives upon the malignity of the whites—but it is true that it lives upon *the voluntary degradation of the blacks.* I do not say that the colored people are more debased than white people would be if persecuted, oppressed and outraged as are the colored people. But I do say that they are debased, deeply debased; and that to recover themselves they must become heroes, self-denying heroes, capable of achieving a great moral victory — a two-fold victory — a victory over themselves and a victory over their enemies."

Here we must close our testimony on this point. The condition of the Free colored people can now be understood. The results, in their case, are vastly different from what was anticipated, when British philanthropists succeeded in West India emancipation. They are very different, also, from what was expected by American Abolitionists — so different, indeed, that their disappointment is fully manifested, in the extracts made from their published documents. As an apology for the failure, it seems to be their aim to create the belief, that the dreadful moral depravation, existing in the West Indies, is wholly owing to the demoralizing tendencies of Slavery. They speak of this effect as resulting from laws inherent in the system, which have no exceptions, and must be equally as active in the United States as in the British colonies. But in their zeal to cast odium on Slavery, they prove too much—for, if this be true, it follows, that the Slave population of the United States must be equally debased with that of Jamaica, and as much disqualified to discharge the duties of freemen, as both have been subjected to the operations of the same system. This is not all. The logic of the argument would extend even to our free colored people, and include them, according to the *American Missionary Association*, in the dire effects of "that enduring legacy which, with its foul, pestilential influences, still blights, like the mildew of death, everything in society that should be lovely, virtuous, and of good report." Now, were it believed, generally, that the colored people of the United States are equally as degraded as those of Jamaica, upon what grounds could any one advocate the admission of the blacks to equal social and political privileges with the whites? Certainly, no Christian family or community would willingly admit such men to terms of social or political equality! This, we repeat, is the logical conclusion from the Reports of the *American Missionary Association* and the *American and Foreign Anti-Slavery Society*—a conclusion, too, the more certain, as it makes no exceptions between the condition of the colored people under the Slavery of Jamaica and under that of the United States.

But in this, as in much connected with Slavery, Abolitionists have taken too limited a view of the subject. They have not properly discriminated between the effects of the original barbarism of the negroes, and the effects produced by the more or less favorable influences to which they were afterward subjected under Slavery. This point deserves special notice. According to the best authorities, the colored people of Jamaica, for nearly three hundred years, were entirely without the Gospel; and it gained a permanent footing among them, only at a few points, at their emancipation, twenty years ago; so that, when liberty reached them, the great mass of the Africans, in the British West Indies, were heathen.* Let us understand the reason of this. Slavery is not an element of human progress, under which the mind necessarily becomes enlightened; but Christianity is the *primary* element of progress, and can elevate

* Rev. Mr. Phillippo, for twenty years a missionary in Jamaica, in his "Jamaica, its Past and Present Condition."

the savage, whether in bondage or in freedom, if its principles are taught him in his youth. The Slavery of Jamaica began with savage men. For three hundred years its slaves were destitute of the Gospel, and their barbarism was left to perpetuate itself. But in the United States, the Africans were brought under the influence of Christianity, on their first introduction, over two hundred and thirty years since, and have continued to enjoy its teachings, in a greater or less degree, to the present moment. The disappearance from among our colored people, of the heathen condition of the human mind—the incapacity to comprehend religious truths—and its continued existence among those of Jamaica, can now be understood. The opportunities enjoyed by the former, for advancement over the latter, have been as *six* to *one*. With these facts before the mind, it is not difficult to perceive, that the colored population of Jamaica, can not but still labor under the disadvantages of *hereditary heathenism* and *involuntary servitude*, with the superadded misfortune of being inadequately supplied with Christian instruction, along with their recent acquisition of freedom. But while all this must be admitted, of the colored people of Jamaica, it is not true of those of our own country; for, long since, they have cast off the heathenism of their fathers, and have become enlightened in a very encouraging degree. Hence it is, that the colored people of the United States, both bond and free, have made vastly greater progress, than those of the British West Indies, in their knowledge of moral duties and the requirements of the Gospel; and hence, too, it is, that GERRITT SMITH is right, in asserting, that the demoralized condition of the great mass of the Free colored people, in our cities, is inexcusable, and deserving of the utmost reprobation, because it is *voluntary*—they knowing their duty, but abandoning themselves to degrading habits.

This brings us to another point of great moment. It will be denied by but few — and by none maintaining the natural equality of the races—that the Free colored people of the United States are sufficiently enlightened, to be elevated, by education, as readily as the whites of similar ages, where equal restraints from vice, and encouragements to virtue prevail. A large portion, even, of the slave population, are similarly enlightened.* We speak not of the state of their morals.

Why is it, then, that the efforts to elevate the Free colored people, have been so unsuccessful? Before answering this question, it is necessary to call attention to the fact, that Abolitionists seem to be sadly disappointed in their expectations, as to the progress of the free colored

---

* As many, at the North, are not aware of the extent to which the religious training of the slaves at the South prevails, we append the following paragraph, in relation to the doings of one denomination, alone, in South Carolina. Similar efforts, more or less extensive, have been made in the other States.

"RELIGIOUS INSTRUCTION OF SLAVES,—The South Carolina Methodist Conference have a missionary committee devoted entirely to promoting the religious instruction of the slave population, which has been in existence twenty-six years. The report of the last year shows a greater degree of activity than is generally known. They have twenty-six missionary stations in which thirty-two missionaries are employed. The report affirms that public opinion in South Carolina is decidedly in favor of the religious instruction of slaves, and that it has become far more general and systematic than formerly. It also claims a great degree of success to have attended the labors of the missionaries."—*N. Y. Evangelist.*

people. Their vexation at the stubbornness of the Negroes, and the consequent failure of their measures, is very clearly manifested in the complaining language, used by GERRITT SMITH, toward the colored people of the eastern cities, as well as by the contempt expressed by the *American Missionary Association*, for the colored preachers of Canada. They had found an apology, for their want of success in the United States, in the presence and influence of Colonizationists; but no such excuse can be made for their want of success in Canada and the West Indies. Having failed in their anticipations, now they would fain shelter themselves under the pretense, that a people once subjected to slavery, even when liberated, can not be elevated in a single generation; that the case of adults, raised in bondage, like heathen of similar age, is hopeless, and their children, only, can make such progress as will repay the missionary for his toil. But they will not be allowed to escape the censure due to their want of discrimination and foresight, by any such plea; as the success of the Republic of Liberia, conducted from infancy to independence, almost wholly by liberated slaves, and those who were born and raised in the midst of Slavery, attests the falsity of their assumption.

But to return. Why have the efforts for the elevation of the free colored people, not been more successful? On this point our remarks may be limited to our own free colored people. The barrier to their progress here, exists not in their want of capacity, but in the absence of the incitements to virtuous action, which are constantly stimulating the white man to press onward and upward in the formation of character and the acquisition of knowledge. There is no position in church or state, to which the poorest white boy, in the common school, may not aspire.

There is no post of honor, in the gift of his country, that is legally beyond his reach. But such encouragements to noble effort, do not reach the colored man, and he remains with us a depressed and disheartened being. Persuading him to remain in this hopeless condition, has been the great error of the Abolitionists. They overlooked the teachings of history, that two races, differing so widely as to prevent their amalgamation by marriage, can never live together, in the same community, but as superiors and inferiors — the inferior remaining subordinate to the superior. The encouraging hopes held out to the colored people, that this law would be inoperative upon them, has led only to disappointment. Happily, this delusion is nearly at an end; and they are beginning to act on their own judgments. They find themselves so scattered and peeled, that there is not another half million of men in the world, so enlightened, who are accomplishing so little for their social and moral advancement. They perceive that they are nothing but branches, wrenched from the great African *banyan*, not yet planted in genial soil, and affording neither shelter nor food to the beasts of the forest or the fowls of the air — their roots unfixed in the earth, and their tender shoots withering as they hang pendent from their boughs.

But little progress, then, it will be seen, has been made, by the free colored people, toward an approximation of equality with the whites.

Have they succeeded better in aiding to abolish Slavery? This question has received its answer in the history of the triumph of Slavery. It is an important one, as this was a principal object influencing them to remain in the country. Their agency in the attempts made to abolish the institution having failed, a more important question arises, as to whether the free colored people, by refusing to emigrate, may not have contributed to the advancement of slavery? An affirmative answer must be given to this inquiry. Nor is a protracted discussion necessary to prove the assertion.

One of the objections urged with the greatest force against Colonization, is, its tendency, as is alleged, to increase the value of slaves by diminishing their numbers. "*Jay's Inquiry*," 1835, presents this objection at length; and the Report of the "*Anti-Slavery Society, of Canada*," 1853, sums it up in a single proposition, thus:

"The first effect of beginning to reduce the number of slaves, by colonization, would be to increase the market value of those left behind, and thereby increase the difficulty of setting them free."

The practical effect of this doctrine, is to discourage all emancipations; to render eternal the bondage of each individual slave, unless all can be liberated; to prevent the benevolence of one master from freeing his slaves, lest his more selfish neighbor should be thereby enriched; and to leave the whole system intact, until its total abolition can be effected. Such philanthropy would leave every individual, of suffering millions, to groan out a miserable existence, because it could not at once effect the deliverance of the whole. This objection to Colonization can be founded only in prejudice, or is designed to mislead the ignorant. The advocates of this doctrine do not practice it, or they would not promote the escape of fugitives to Canada.

But Abolitionists object not only to the Colonization of liberated slaves, as tending to perpetuate Slavery; they are equally hostile to the Colonization of the Free colored people, for the same reason. The "*American Reform Tract and Book Society*," the organ of the Abolitionists, for the publication of Anti-Slavery works, has issued a Tract on "Colonization," in which this objection is stated as follows:

"The Society perpetuates Slavery, by removing the free laborer, and thereby increasing the demand for, and the value of, Slave labor."

The projectors and advocates of such views may be good philanthropists, but they are bad philosophers. We have seen that the power of American Slavery, lies in the demand for its products; and that the whole country, north of the sugar and cotton States, is actively employed in the production of provisions for the support of the planter and his slaves, and in consuming the products of Slave labor. This is the constant vocation of the whites. And how is it with the blacks? Are they competing with the slaves, in the cultivation of sugar and cotton, or are they also supporting the system, by consuming its products? The latitudes in which they reside, and the pursuits in which they are engaged, will answer this question.

The census of 1850, shows but 40,900 free

colored persons in the nine sugar and cotton States, including Texas, Louisiana, Arkansas, Tennessee, Mississippi, Alabama, Georgia, Florida, and South Carolina, while 393,500 are living in the other States. North Carolina is omitted, because it is more of a tobacco and wool-growing, than cotton-producing State. Of the free colored persons in the first-named States, 19,260 are in the cities and larger towns; while, of the remainder, a considerable number may be in the villages, or in the families of the whites. From these facts it is apparent, that less than 20,000 of the entire Free colored population (omitting those of North Carolina), are in a position to compete with Slave labor, while all the remainder, numbering over 412,800, are engaged, either directly or indirectly, in supporting the institution. Even the fugitives escaping to Canada, from having been producers necessarily become consumers of Slave-grown products; and, worse still, under the reciprocity treaty, they must also become growers of provisions for the planters who continue to hold their brothers, sisters, wives and children, in bondage.

These are the practical results of the policy of the Abolitionists. Verily, they, also, have dug their ditches on the wrong side of their breastworks, and afforded the enemy an easy entrance into their fortress. But, "Let them alone; they be blind leaders of the blind. And if the blind lead the blind, both shall fall into the ditch."*

But a brighter day is dawning for the Free colored people. They are wearied in watching for the "better time coming," promised by their white friends, and are unwilling to "wait a little longer," as runs one of their songs of inaction. To collect their scattered fragments; to consolidate their divided forces; to sink their individual popularity into an honored nationality, is now the aim of their thoughtful men.

But where is this great achievement to be made? Not in the organization of a new government, as no part of the earth remains unoccupied. It must be by a fusion with one already established. But what one? Not with one like the British Colonies, in subjection to a distant

*Matthew's Gospel, xv, 14.

throne, and nearly destitute of schools and all the means of intellectual and moral improvement. It must be with one possessing the elements of progress—which offers peace, security, prosperity, liberty, equality, fraternity, and Protestant Christianity. No other will meet their wants; nor should any other be adopted, as worthy colored freemen, who have caught the spirit of the republican institutions of the United States. South America can afford no suitable asylum, as the diversity of language, and the antagonism of its religion, together with the frequency of its civil wars, and the insecurity of property and life, forbid their choosing a home in that region.

Thus, Liberia is the only nation with which a fusion, by the free colored people, can be safely made. While remaining here, they must continue to support Slavery, and suffer from inadequate means of improvement. The only portion of their number, who have escaped from all connection with Slavery, are those who have removed to Liberia. In that Republic, too, all the necessary stimulants to civil, social, intellectual, and moral advancement, are within the reach of the colored man. Nor are they left to the contingencies of the varying prosperity or adversity of the colonists, for their perpetuation. The four great leading Churches in the United States—the Episcopal, the Methodist, the Presbyterian, and the Baptist—are pledged to the support of its educational and religious institutions; and hence, while generations will certainly be needed for the elevation of the Free colored people here, strive as they may, a single one, with right-hearted men, can do the work there.

4. The moral relations of persons holding the *per se* doctrine, on the subject of Slavery, to the purchase and consumption of Slave-labor products.

Having noticed the political and economical relations of Slavery, it may be expected that we shall say something of its moral relations. In attempting this, we choose not to traverse that interminable labyrinth, without a thread, which includes the moral character of the system, as it respects *the relation between the master and his slave.* Such questions are left for those who believe themselves skilled in resolving cases of conscience, or framing terms of Church communion. The only aspect in which we care to consider it, is in *the moral relations which the consumers of Slave-labor products sustain to Slavery*— and even on this we shall offer no opinion, our aim being only to promote inquiry.

This view of the question is not an unimportant one. It includes the germ of the grand error in nearly all Anti-Slavery effort; and to which, chiefly, is to be attributed its want of moral power over the conscience of the Slaveholder. The recent Abolition movement, was designed to create a public sentiment, in the United States, that should be equally as potent in forcing emancipation, as was the public opinion of Great Britain. But why have not the Americans been as successful as the English? This is an inquiry of great importance. When the Anti-Slavery Convention, which met, December 6, 1833, in Philadelphia, declared, as a part of its creed: "That there is no difference in principle, between the African Slave Trade, and American Slavery," it meant to be understood as teaching, that persons who purchased slaves imported from Africa, or who held their offspring as slaves, were *particeps criminis*, partakers in the crime, with the Slave-trader, on the principle that he who receives stolen property, knowing it to be such, is equally guilty with the thief.

On this point DANIEL O'CONNELL was very explicit, when, in a public assembly, he used this language: "When an American comes into Society, he will be asked, 'are you one of the thieves, or are you an honest man? If you are an honest man, then you have given liberty to your slaves; if you are among the thieves, the sooner you take the outside of the house, the better.'"

The error just referred to was this: they based their opposition to Slavery on the principle, that it was *malum in se*, a sin in itself, like the Slave trade, robbery, and murder; and, at the same time, continued to use the products of the labor of the slave as though they had been obtained from the labor of freemen. But this seeming inconsistency, was not the only reason why they failed to create such a public sentiment, as would procure the emancipation of our slaves. The English Emancipationists began their work like philosophers—addressing themselves respectfully, to the power that could grant their requests. Beside the moral argument, which declared Slavery a crime, the English philanthropists labored to convince Parliament, that emancipation would be advantageous to the commerce of the nation. The commercial value of the Islands had been reduced one-third, as a result of the Abolition of the Slave trade. Emancipation, it was argued, would more than restore their former prosperity, as the labor of freemen was twice as productive as that of slaves. But American Abolitionists commenced their crusade against Slavery, by charging those who sustained it, and who alone, held the power to manumit, with crimes of the blackest die. This placed the parties in instant antagonism, causing all the arguments on human rights, and the sinfulness of Slavery, to fall without effect upon the ears of angry men. The error on this point, consisted in failing to discriminate between the sources of the power over emancipation in England and in the United States. With Great Britain, the power was in Parliament. The masters, in the West Indies, had no voice in the question. It was the voters in England alone who controlled the elections, and, consequently, controlled Parliament. But the condition of things in the United States is the reverse of what it was in England. With us, the power of emancipation is in the States, not in Congress. The Slaveholders elect the members to the State Legislatures; and they choose none but such as agree with them in opinion. It matters not, therefore, what public sentiment may be at the North, as it has no power over the Legislatures of the South. Here, then, is the difference: with us the Slaveholder controls the question of emancipation—in England the consent of the master was not necessary to the execution of that work.

Our Anti-Slavery men seem to have fallen into their errors of policy, by following the lead of those of England, who manifested a total ignorance of the relations existing between our

General Government and that of the States. On the Abolition platform, Slaveholders found themselves placed in the same category with Slave-traders and thieves. They were told, that all laws giving them power over the slave, were void, in the sight of Heaven; and that their appropriation of the fruits of the labor of the slave, was robbery. Had the preaching of these principles produced conviction, it must have promoted emancipation. But, unfortunately, while these doctrines were held up to the gaze of Slaveholders, in the one hand of the exhorter, they beheld his other hand stretched out, from beneath his cloak of seeming sanctity, to clutch the products of the very robbery he was professing to condemn! Take a fact in proof of this view of the subject.

At the date of the declarations of DANIEL O'CONNELL, on behalf of the English, and by the Philadelphia Anti-Slavery Convention, on the part of Americans, the British Manufacturers were purchasing, annually, about 300,000,000 lbs. of cotton, from the very men denounced as equally criminal with Slave-traders and thieves; and the people of the United States were almost wholly dependent upon Slave labor for their supplies of cottons and groceries. It is no matter for wonder, therefore, that Slaveholders should treat, as fiction, the doctrine that Slave-labor products are the fruits of robbery, so long as they are purchased, without scruple, by all classes of men, in Europe and America. The pecuniary argument for emancipation, that Free labor is more profitable than Slave labor, was also urged here; but was treated as the greatest absurdity. The masters had before their eyes, the evidence of the falsity of the assertion, that, if emancipated, the Slaves would be doubly profitable as free laborers. The reverse was admitted, on all hands, to be true in relation to our colored people.

But this question, of the moral relations which the consumers of slave-labor products sustain to Slavery, is one of too important a nature to be passed over without a closer examination; and, beside, it is involved in less obscurity than the morality of the relation existing between the master and the slave. Its consideration, too, affords an opportunity of discriminating between the different opinions entertained on the broad question of the morality of the institution, and enables us to judge of the consistency and consciousness of every man, by the standard which he himself adopts.

The prevalent opinions, as to the morality of the Institution of Slavery, in the United States, may be classified under three heads. 1. That it is justified by Scripture example and precept. 2. That it is a great civil and social evil, resulting from ignorance and degradation, like despotic systems of Government, and may be tolerated until its subjects are sufficiently enlightened to render it safe to grant them equal rights. 3. That it is *malum in se,* a sin in itself, like robbery and murder, and cannot be sustained, for a moment, without sin; and, like sin, should be immediately abandoned.

Those who consider Slavery sanctioned by the Bible, conceive that they can, consistently with their creed, not only hold slaves, and use the products of slave labor, without doing violence to their consciences, but may adopt measures to perpetuate the system. Those who consider Slavery merely a great civil and social evil, a despotism that may engender oppression, or may not, are of opinion that they may purchase and use its products, or interchange their own for those of the Slaveholder, as free Governments hold commercial and diplomatic intercourse with despotic ones, without being responsible for the moral evils connected with the system. But the position of those who believe Slavery *malum in se*, like the slave trade, robbery and murder, is a very different one from either of the other classes, as it regards the purchase and use of Slave-labor products. Let us illustrate this by a case in point:

A company of men hold a number of their fellow-men in bondage, under the laws of the commonwealth in which they live, so that they can compel them to work their plantations, and raise horses, cattle, hogs, and cotton. These products of the labor of the oppressed, are appropriated by the oppressors to their own use, and taken into the markets for sale. Another company proceed to a community of freemen, who have labored voluntarily during the year, seize their persons, bind them, convey away their horses, cattle, hogs, and cotton, and take the property to market. The first association represents the Slaveholders; the second a band of robbers. The commodities of both parties, are openly offered for sale, and every one knows how the property of each was obtained. Those who believe the *per se* doctrine, place both these associations in the same moral category, and call them robbers. Judged by this rule, the first band are the more criminal, as they have deprived their victims of personal liberty, forced them into servitude, and then taken from them, without pay, the proceeds of their labor. The second band have only deprived their victims of liberty, while they robbed them; and thus have committed but two crimes, while the first have perpetrated three. These parties attempt to negotiate the sale of their cotton, say in London. The first company dispose of their cargo without difficulty — no one manifesting the slightest scruple at purchasing the products of Slave-labor. But the second company are not so fortunate. As soon as their true character is ascertained, the police drag its members to Court, where they are sentenced to Bridewell. In vain do these robbers quote the Philadelphia Anti-Slavery Convention, and Daniel O'Connell, to prove that their cotton was obtained by means no more criminal than that of the Slaveholders, and that, therefore, judgment ought to be reversed. The Court will not entertain such a plea, and they have to endure the penalty of the law. Now, why this difference, if Slavery be *malum in se?* And if the receiver of stolen property is *particeps criminis* with the thief, why is it, that the Englishman, who should receive and sell the cotton of the robbers, would run the risk of being sent to prison with them, while if he acted as agent of the Slaveholders, he would be treated as an honorable man? If the master has no moral right to hold his slaves, in what respect can the products of their labor differ from the property acquired by robbery? And if the property be the fruits of robbery, how can any one use it, without violating conscience?

We have met with the following sage exposition of the question, in justification of the use of Slave-labor products, by those who believe the *per se* doctrine: The master owns the lands, gives his skill and intelligence to direct the labor, and feeds and clothes the slaves. The slaves, therefore, are entitled only to a part of the proceeds of their labor, while the master is also justly entitled to a part of the crop. When brought into market, the purchaser can not know what part belongs, rightfully, to the master and what to his slaves, as the whole is offered in bulk. He may, therefore, purchase the whole, innocently, and throw the sinfulness of the transaction upon the master, who sells what belongs to others. But if the *per se* doctrine be true, this apology for the purchaser, is not a justification. Where a "confusion of goods" has been made by one of the owners, so that they cannot be separated, he who "confused" them can have no advantage, in law, from his own wrong, but the goods are awarded to the innocent party. On this well known principle of law, this most equitable rule, the master forfeits his right in the property, and the purchaser, knowing the facts, becomes a party in his guilt. But aside from this, the "confusion of goods," by the master, can give him no moral right to dispose of the interest of his slaves therein for his own benefit; and the persons purchasing such property, acquire no moral right to its possession and use. These are sound, logical views. The argument offered, in justification of those who hold that Slavery is *malum in se*, is the strongest that can be made. It is apparent, then, from a fair analysis of their own principles, that they are *particeps criminis* with Slaveholders.

Again, if the laws regulating the institution of Slavery, be morally null and void, and not binding on the conscience, then, the slaves have a moral right to the proceeds of their labor. This right can not be alienated by any act of the master, but attaches to the property wherever it may be taken, and to whomsoever it may be sold. This principle, in law, is also well established. The recent decision on the "Gardiner fraud," confirms it; the Court asserting, that the money paid out of the Treasury of the United States, under such circumstances, continued its character as the money and property of the United States, and may be followed into the hands of those who cashed the orders of Gardiner, and subsequently drew the money, but who are not the true owners of the said fund; and decreeing that the amount of funds, thus obtained, be collected off the estate of said Gardiner, and off those who drew funds from the Treasury, on his orders.

These principles of law are so well understood, by every man of intelligence, that we can not conceive how those advocating the *per se* doctrines, if sincere, can continue in the constant use of Slave-grown products, without a perpetual violation of conscience and of all moral law. Taking them under *protest*, against the Slavery which produced them, is ridiculous. Refusing to fellowship the Slaveholer, while eagerly appropriating the products of the labor of the slave, which he brings in his hand, is contemptible. The most noted case of the kind, is that of the British Committee, who had charge of the preliminary arrangements for the admission of members to the WORLD'S CHRISTIAN EVANGELICAL ALLIANCE. One of the rules it adopted, but which the Alliance afterward modified, excluded all American clergymen, suspected of a want of orthodoxy on the *per se* doctrine, from seats in that body. Their language, to American clergymen, was virtually, "Stand aside, I am holier than thou; while, at the same moment their parishioners, the manufacturers, had about completed the purchase of 624,000,000 lbs. of cotton, for the consumption of their mills, during the year; the bales of which, piled together, would have reached mountain-high, displaying mostly the brands, "New Orleans," "Mobile," "Charleston."

As not a word was said, by the Committee, against the Englishmen who were buying and manufacturing American cotton—the case may be viewed as one in which the fruits of robbery were taken under *protest* against the robbers themselves. To all intelligent men, the conduct of the people of Britain, in protesting against Slavery, as a system of robbery, while continuing to purchase such enormous quantities of the cotton produced by slaves, appears as Pharisaical as the conduct of the *conscientious* Scotchman, in early times, in Eastern Pennsylvania, who married his wife under protest against the Constitution and laws of the Government, and, especially, against the authority, power, and right of the magistrate who had just tied the knot.*

* An anecdote, illustrative of the pliability of some consciences, of this apparently rigid class, where interest or inclination demands it, has often been told by the late Governor Morrow, of Ohio. An old Scotch "Cameronian," in Eastern Pennsylvania, became a widower, shortly after the adoption of the Constitution of the United States. He refused to acknowledge either the National or State Governments, but pronounced them both unlawful, unrighteous, and ungodly. Soon he began to feel the want of a wife, to care for his motherless children. The consent of a woman in his own church was gained, because to take any other would have been like an Israelite marrying a daughter of the land of Canaan. On this point, as in refusing to swear allegiance to Government, he was controlled by conscience. But now a practical difficulty presented itself. There was no minister of his church in the country—and those of other denominations, in his judgment, had no Divine warrant for exercising the functions of the sacred

Such pliable consciences, doubtless, are very convenient in cases of emergency. But as they relax when selfish ends are to be subserved, and

---

office. He repudiated the whole of them. But how to get married, that was the problem. He tried to persuade his intended to agree to a marriage contract, before witnesses, which could be confirmed whenever a proper minister should arrive from Scotland. But his "lady-love" would not consent to the plan. She must be married "like other folk," or not at all—because "people would talk so." The Scotchman for want of a wife, like Great Britain for want of cotton, saw very plainly that his children must suffer; and so he resolved to get married, at all hazards, as England buys her cotton, but so as not to violate conscience. Proceeding, with his intended, to a magistrate's office, the ceremony was soon performed, and they twain pronounced "one flesh." But no sooner had he "kissed the bride," the sealing act of the contract at that day, than the good Cameronian drew a written document from his pocket, which he read aloud, before the officer and witnesses; and in which he entered his solemn protest against the authority of the Government of the United States, against that of the State of Pennsylvania, and especially against the power, right, and lawfulness of the acts of the magistrate, who had just married him. This done, he went his way, rejoicing that he had secured a wife without recognizing the lawfulness of ungodly Governments, or violating his conscience.

retain their rigidity only when judging the conduct of others, the inference is, that the persons possessing them are either hypocritical, or else, as was acknowledged by Parson D., in similar circumstances, they have mistaken their *prejudices* for their *consciences.*

So far as Britain is concerned, she is, manifestly, much more willing to receive American Slave-labor cotton for her factories, than American republican principles for her people. And why so? The profits derived by her, from the purchase and manufacture of Slave-labor cotton, constitute so large a portion of the means of her prosperity, that the government could not sustain itself were the supplies of this article cut off. It is easy to divine, therefore, why the people of England are boundless in their denunciations of American Slavery, while not a single remonstrance goes up to the throne, against the import of American cotton. Should she exclude it, the act would render her unable to pay the interest on her national debt; and many a declaimer against Slavery, losing his income, would have to go supperless to bed.

Let us contrast the conduct of a pagan government with that of Great Britain. When the Emperor of China became fully convinced of his inability to resist the prowess of the British arms, in the famous "Opium War," efforts were made to induce him to legalize the traffic in opium, by levying a duty on its import, that should yield him a heavy profit. This he refused to do, and recorded his decision in these memorable words:

"It is true I can not prevent the introduction of the flowing poison. Gain-seeking and corrupt men will, for profit and sensuality, defeat my wishes, but nothing will induce me to derive a revenue from the vice and misery of my people."*

The reason can now be clearly comprehended, why Abolitionists have had so little moral power over the conscience of the Slaveholder. Their practice has been inconsistent with their precepts; or, at least, their conduct has been liable to this construction. Nor do we perceive how they can exert a more potent influence, in the

* National Intelligencer, 1854.

future, unless their energies are directed to efforts such as will relieve them from a position so inconsistent with their professions, as that of constantly purchasing products which they, themselves, declare to be the fruits of robbery. While, therefore, things remain as they are, with the world so largely dependent upon Slave labor, how can it be otherwise, than that the system will continue to flourish? And while its products are used by all classes, of every sentiment, and country, nearly, how can the Slaveholder be brought to see anything, in the practice of the world, to alarm his conscience, and make him cringe, before his fellow-men, as a guilty robber?

But, has nothing worse occurred from the advocacy of the *per se* doctrine, than an exhibition of inconsistency on the part of Abolitionists, and the perpetuation of Slavery resulting from their conduct? This has occurred. Three highly respectable religious denominations, now limited to the North — the Associate Presbyterian Church, the Reformed Presbyterian Church, and the Associate Reformed Presbyterian Church—had once many flourishing congregations in the South. On the adoption of the *per se* doctrine, by their respective Synods, these congregations became disturbed, were soon after broken up, or the ministers in charge had to seek other fields of labor. Their system of religious instruction, for the family, being quite thorough, the Slaves were deriving much advantage from the influence of these bodies. But when they resolved to withhold the Gospel from the Master, unless he would emancipate, they also withdrew the means of grace from the Slave; and, so far as they were concerned, left him to perish eternally! Whether this course was proper, or whether it would have been better to have passed by the morality of the legal relation, in the creation of which the master had no agency, and considered him, under Providence, as the moral guardian of the Slave, bound to discharge a guardian's duty to an immortal being, we shall not undertake to determine. Attention is called to the facts, merely, to show the practical effects of the action of these Churches upon the Slave, and what the *per se* doctrine has done in depriving him of the Gospel.

Another remark, and we have done with this topic. Nothing is more common, in certain circles, than denunciations of the Christian men and ministers, who refuse to adopt the *per se* principle. We leave others to judge whether these censures are merited. One thing is certain: those who believe that Slavery is a great civil and social evil, entailed upon the country, and are extending the Gospel to both Master and Slave, with the hope of removing it peacefully, cannot be reproached with acting inconsistently with their own principles; while those who declare Slavery *malum in se*, and refuse to fellowship the Christian Slaveholder, but yet use the products of Slave-labor, may fairly be classified, on their principles, with the hypocritical people of Israel, who were thus reproached by the Most High: "What hast thou to do to declare my statutes, or that thou shouldst take my covenant in thy mouth? * * * When thou sawest a thief, then thou consentedst with him."*

---

* Psalm l, 16, 18.

CONCLUSION.

In concluding our labors, there is little need of extended observation. The work of Emancipation, in our country, was checked, and the extension of Slavery promoted:—first, by the Free Colored People neglecting to improve the advantages afforded them; second, by the increasing value imparted to Slave-labor; third, by the mistaken policy into which the Abolitionists have fallen. Whatever reasons might now be offered, for emancipation, from an improvement of our Free colored people, is far more than counterbalanced by its failure in the West Indies, and the constantly increasing value of the labor of the Slave. If, when the Planters had only a moiety of the markets for Cotton, the value of Slavery was such as to arrest emancipation, how must the obstacles be increased, now, when they have the monopoly of the markets of the world?

We propose not to speak of remedies for Slavery. That we leave to others. Thus far this great civil and social evil, has baffled all human wisdom. Either some radical defect must have existed, in the measures devised for its removal, or the time has not yet come for successfully assailing the Institution. Our work is completed, in the delineation we have given of its varied relations to our commercial and social interests. As the monopoly of the culture of Cotton, imparts to Slavery its economical value, the system will continue as long as this monopoly is maintained. Slave-Labor products have now become necessities of human life, to the extent of more than half the commercial articles supplied to the Christian world. Even Free labor, itself, is made largely subservient to Slavery, and vitally interested in its perpetuation and extension.

Can this condition of things be changed? It may be reasonably doubted, whether anything efficient can be speedily accomplished: not because there is lack of territory where freemen may be employed in tropical cultivation; not because intelligent free-labor is less productive than slave-labor; but because freemen, whose constitutions are adapted to tropical climates, will not avail themselves of the opportunity offered for commencing such an enterprise.

King Cotton cares not whether he employs slaves or freemen. It is the *cotton*, not the *slaves*, upon which his throne is based. Let freemen do his work as well, and he will not object to the change. Thus far the experiments in this respect have failed, and they will not soon be renewed. The efforts of his most powerful ally, Great Britain, to promote that object, have already cost her people many hundreds of millions of dollars; with total failure as a reward for her zeal. One-sixth of the colored people of the United States are free; *but they shun the cotton regions*, and have been instructed to detest *emigration to Liberia.* Their improvement has not been such as was anticipated; and their more rapid advancement cannot be expected, while they remain in the country. The free colored people of the West Indies, can no longer be relied on to furnish tropical products, for they are fast sinking into savage indolence. His Majesty, King Cotton, therefore, is forced to continue the employment of his slaves; and, by their toil, is riding on, conquering and to conquer! He receives no check from the cries of the oppressed, while the citizens of the world are dragging forward his chariot, and shouting aloud his praise!

KING COTTON is a profound statesman, and knows what measures will best sustain his throne. He is an acute mental philosopher, acquainted with the secret springs of human action, and accurately perceives who will best promote his aims. He has no evidence that colored men can grow his cotton, but in the capacity of slaves. It is his policy, therefore, to defeat all schemes of emancipation. To do this, he stirs up such agitations as lure his enemies into measures that will do him no injury. The venal politician is always at his call, and assumes the form of saint or sinner, as the service may demand. Nor does he overlook the enthusiast, engaged in Quixotic endeavors for the relief of suffering humanity, but influences him to advocate measures which tend to tighten, instead of loosing the bands of Slavery. Or, if he cannot be seduced into the support of such schemes, he is beguiled into efforts that waste his strength on objects the most impracticable — so that Slavery receives no damage from the exuberance of his philanthropy. But should such a one, perceiving the futility of his labors, and the evils of his course, make an attempt to avert the consequences; while he is doing this, some new recruit, pushed forward into his former place, charges him with lukewarmness, or Pro-slavery sentiments, destroys his influence with the public, keeps alive the delusions, and sustains the supremacy of KING COTTON in the world.

In speaking of the economical connections of Slavery with the other material interests of the world, we have called it a *tri-partite alliance.* It is more than this. It is *quadruple.* Its structure includes four parties, arranged thus: The Western Agriculturists; the Southern Planters; the English Manufacturers; and the American Abolitionists! By this arrangement, the Abolitionists do not stand in direct contact with Slavery:— they imagine, therefore, that they have clean hands and pure hearts, so far as sustaining the system is concerned But they, no less than their allies, aid in promoting the interests of Slavery. Their sympathies are with England on the Slavery question, and they very naturally incline to agree with her on other points. She advocates *Free Trade*, as essential to her manufactures and commerce; and they do the same, not waiting to inquire into its bearings upon *American Slavery*. We refer now to the people, not to their leaders, whose integrity we choose not to indorse. The Free Trade and Protective Systems, in their bearings upon Slavery, are so well understood, that no man of general reading, especially an editor, who professes Anti-Slavery sentiments, at the same time advocating Free Trade, will ever convince men of intelligence, pretend what he may, that he is not either woefully perverted in his judgment, or emphatically, a "dough-face" in disguise! England, we were about to say, is in alliance with the cotton planter, to whose prosperity Free Trade is indispensable. Abolitionism is in alliance with England. All three of these parties, then, agree in their support of the Free Trade policy. It needed but the aid of the Western

Farmer, therefore, to give permanency to this principle. His adhesion has been given, the *quadruple alliance* has been perfected, and Slavery and Free Trade *nationalized!*

The crisis now upon the country, as a consequence of Slavery having become dominant, demands that the highest wisdom should be brought to the management of national affairs. The *quacks* who have aided in producing the malady, and who have the effrontery still to claim the right to manage the case, must be dismissed. The men who mock at the Political Economy of the North, and have assisted in crushing its cherished policy, must be rebuked. Slavery, *nationalized,* can now be managed only as a national concern. It can now be abolished only with the consent of those who sustain it. Their assent can be gained only on employing other agents to meet the wants it now supplies. It must be superseded, then, if at all, by means that will not injuriously affect the interests of commerce and agriculture, to which it is now so important an auxiliary. To supply the demand for tropical products, except by the present mode, is not the work of a day, nor of a generation. Should the influx of foreigners continue, such a change may be possible. But to effect the transition from Slavery to Freedom, on principles that will be acceptable to the parties who control the question; to devise and successfully sustain such measures as will produce this result; must be left to statesmen of broader views and loftier conceptions than are to be found among those at present engaged in this great controversy.

In noticing the *strategy* by which the Abolitionists were rendered subservient to Slavery, through the ignorance or duplicity of their leaders, we refer to the political action, only, in which they were induced to participate. We yield to none in our veneration for the early Anti-Slavery men, whose zeal for the overthrow of oppression, and the relief of the country from its greatest curse, was kindled at the altar of a pure philanthropy; and to whom official honors and emoluments had few attractions. We intend not to disparage such men.

Those who believe that Slavery is a *divine*

institution, which should be perpetuated; as well as those who hold the sentiment, that it is a *malum in se*, that must be instantly abandoned; entertain views so much at variance with the practical judgment of the world, that they can never hope to see their principles become dominant. The doctrine of the *divine right of Slavery*, is as repugnant to the spirit of the age, as that of the *divine right of kings, or of popes.* The *per se* doctrine, more plausible at first view, is everywhere practically repudiated, in the business transactions of the world; and involves those who profess it, not only in every-day inconsistencies, but bars their access to the master and dooms the slave to perpetual ignorance.

These two extreme views can not become prevalent; but must remain circumscribed within the narrow limits to which they have been hitherto confined. It is well for the country that it is so. These parties are so antagonistic, that their policy has harmonized in nothing but the triumph of Slavery, and the increase of the dangers of a dissolution of the Union.

The view, that Slavery is a great *civil and*

social evil, identical in principle with despotism, is beset with fewer difficulties, meets with less opposition, and is likely to become the prevalent belief of the world. This view maintains, that Slavery is an incubus, pressing on humanity, like despotism in any other form; and *sinful,* only, so far as it abuses its power. This liability to abuse, it is admitted, is increased under American Slavery, from the fact, that while a single despot often governs many millions of subjects, with us, three hundred and fifty thousand masters rule over but three millions two hundred and fifty thousand slaves: subjecting them, not to uniform laws, but to an endless diversity of treatment, as benevolence or cupidity may dictate.

How far masters in general escape the commission of sin, in the treatment of their slaves, or whether any are free from guilt, is not the point at issue, in this view of Slavery. The mere possession of power over the slave, under the sanction of law, is held not to be sinful; but, like despotism, may be used for the good of the governed. Here arises a question of importance:

Can despotism be acknowledged, by Christians, as a lawful form of government? Those who hold the view of slavery under consideration, answer in the affirmative. The necessity of civil government, they say, is denied by none. Society can not exist in its absence. Republicanism can be sustained only where the majority are intelligent and moral. In no other condition can free government be maintained. Hence, despotism establishes itself, of necessity, more or less absolutely, over an ignorant or depraved people; obtaining the acquiescence of the enlightened, by offering them security to person and property. Few nations, indeed, possess moral elevation sufficient to maintain republicanism. Many have tried it; have failed, and relapsed into despotism. Republican nations, therefore, must either forego all intercourse with despotic governments, or acknowledge them to be lawful. This can be done, it is claimed, without being accountable for moral evils connected with their administration. Elevated examples of such recognitions are on record. Christ paid tribute to Cæsar; and Paul admitted the validity of the despotic government of Rome, with its thirty millions of slaves. To deny the lawfulness of despotism, and yet hold intercourse with such governments, is as inconsistent as to hold the *per se* doctrine, in regard to Slavery, and still continue to use its products. Slavery and despotism being identical in principle, it follows, that the considerations which justify the recognition of the one, will apply equally to the other.

Another thought, in this connection, crowds itself upon the attention, and demands a hearing. Despotism, though recognized as lawful, from necessity, is repugnant to enlightened and moral men. The notions of equity, everywhere prevailing, makes them revolt at the idea of despotism continuing perpetually. But continue it will, in one form or another, until ignorance is banished, and the moral elevation of mankind effected. Hence it is, that Christian philanthropists, clearly comprehending the truth on this point, have labored, unremittingly, from the days of JOHN KNOX, the Scotch reformer, to the present moment, to promote education among the people, and thus prepare them for the enjoyment of civil liberty. Every consideration, leading Christian men to labor to supersede despotism by republicanism, demands, with equal force, that Slavery shall be superseded by Freedom. There is an advantage gained, it is thought, in ranking Slavery and despotism as identical. It links the fate of the one with that of the other. None but fanatics, however, will attempt to reap before they sow. None who comprehend the causes of the failure of republicanism in France, and of emancipation in Hayti and Jamaica, will desire to witness a repetition of the tragedies there enacted. The benefits repaid not the treasure and the blood they cost. But these tragedies have taught a lesson easily comprehended. Moral elevation must precede the enjoyment of civil privileges. The advance in the former, must be the measure by which to regulate the grant of the latter; otherwise the safety of society is endangered. Upon these principles most of the States have acted, in denying to the Free colored people an equality of political rights. It is a conviction of this truth, that now agitates the public mind, on the question of limiting the political privileges of foreigners, who may hereafter ask the rights of citizenship; and begets the hostility, among Americans, to excluding the Bible from Common Schools. But why so much zeal, it is asked, for the Bible in Common Schools? In the language of another, we, in turn, would ask:

"How comes it that that little volume, composed by humble men in a rude age, when art and science were but in their childhood, has exerted more influence on the human mind and on the social system, than all the other books put together? Whence comes it that this book has achieved such marvelous changes in the opinions of mankind — has banished idol worship — has abolished infanticide — has put down polygamy and divorce — exalted the condition of woman — raised the standard of public morality — created for families that blessed thing, a Christian home — and produced its other triumphs by causing benevolent institutions, open and expansive, to spring up as with the wand of enchantment? What sort of a book is this, that even the winds and waves of human passion obey it? What other engine of social improvement has operated so long, and yet lost none of its virtues? Since it appeared, many boasted plans of amelioration have been tried and failed, many codes of jurisprudence have arisen, and run their course, and expired. Empire after empire has been launched upon the tide of time, and gone down, leaving no trace upon the waters. But this book is still going about doing good, leaving with society its holy principles — cheering the sorrowful with its consolation — strengthening the tempted — encouraging the patient — calming the troubled spirit — and smoothing the pillow of death. Can such a book be the offspring of human genius? Does not the vastness of its effects demonstrate the excellency of the power to be of God?"

The feeling of every true American, on this question, may be thus expressed: "Rather than have my offspring deprived of free access to the fountain of all true morality — rather than see the children of my country deprived of the Bible — I would sacrifice all to prevent such a calamity. With the banishment of the Bible from common-schools, farewell to republicanism—farewell to morality—farewell to religion!"

It is matter of rejoicing, to all who hold these sentiments, that the work of instruction, among the slaves, under the supervision of several of the largest religious denominations in the country, is progressing, slowly, it may be, but successfully. The Bible is among the slaves as well as the masters. The presence of the missionary, engaged in his labor of love, in the midst of the slave population, is an ample demonstration, that the master recognizes his slave as an immortal being, with a soul to be saved or lost. With this work of instruction, increased and perpetuated, the slave will, one day, reach that point of moral elevation, when his bondage may be safely superseded by freedom.

But what of the Free colored people? Their condition and prospects are before the reader. Their agency in checking emancipation, when it was in successful progress, has become history. Their submission, voluntarily, to become "hewers of wood and drawers of water," is a melancholy fact, visible to all. Whoever projects a practicable scheme of abolition, that will again offer inducements to general emancipation, and hasten the redemption of the colored race, must include in his measures, as the first and radical principle, the elevation of those already free! Accomplish this, and more than half the work is completed. The theater for such an achievment is not the United States. It is Africa — Liberia! *Utopia* is not the field — it must be abandoned. Christian men at the South, now hesitate to emancipate their slaves, and cast them, helpless, upon the frigid charities of the North! But let Africa be once redeemed, let civilization and Christianity spread over a few millions of its population, and the moral effect would be irresistible. Every rational objection to emancipation would be at an end. Every Christian master, as his slaves attained sufficient moral elevation, would say to them, "Brothers, go free!"

APPENDIX.

TABLE I.

Facts in relation to Cotton—its growth, manufacture, and influence on Commerce, Slavery, Emancipation, etc., chronologically arranged.

YEARS.	*Gr. Brit. Ann. Imp. & Consump. of Cotton, from earliest dates to 1853, in lbs.*	*United States' Annual Exports Cotton to Gr. Britain and Europe generally.*	*Great Britain's sources of Cotton supplies other than the U. States, with total Cotton crop of United States.*	*Dates of Inventions promoting the growth and manufacture of Cotton, and of movements to elevate the African race.*
1641	Cotton manufac. first named in English history. TOTAL IMPORTS		Previous to 1791 Great Britain obtained her supplies of Cotton from the W. Indies, S. America, and the countries around the eastern parts of the Mediterranean.	Previous to the invention of the machinery named below, all *carding, spinning, and weaving* of wool and cotton had been done by the use of the *hand-cards, one-spindle wheels*, and common *hand-looms*. The work, for a long period, was performed in families; but the improved machinery propelled by steam power has so reduced the cost of cotton manufactures, that all household manufacturing has long since been abandoned, and the monopoly yielded to capitalists, who now fill the world with their cheap fabrics.
1697	1,976,359			
1701	1,985,868			
1700 to 1705	av. 1,170,881			
1710	715,008			
1720	1,972,805			
1730	1,545,472			
1741	1,645,031	(1747–48,) 7 bags of Cotton were shipped fr'm Charleston, S. C., to England.		(1762) *Carding machine* invented. (1767) *Spinning Jenny* invented.
1751	2,976,610			(1769) *Spinning Roller-Frame* invented.
1764	3,870,392			(1769) Cotton first planted in U. States. (") *Watts' Steam Engine* patented.

Year				
1771 to 1775	*av.* 6,766,613	(1770) 2,000 lbs. shipped fr. Charleston.	. . . .	*Mule Jenny* invented. (1776) Virginia forbids foreign slave trade. (1780) Emancipation by Penn. and Mass. (1781) *Muslins* first made in England. (1784) Emancipation by Conn. and R. Isl'd. (1785) *Watts'* Engine improved and applied to cotton machinery. First mill erected 1783. (1785) *N. York Abolition Soc.* organized.
1781	5,198,778			
1782	11,828,039			
1783	9,735,663			
1784	11,482,083	71 bags shipped and seized in England, on the ground that America could not produce so much		
1785	18,400,384			
1786	19,475,020		*Imports* by Great Britain from Br. West Indies, lbs. 5,800,000; Fr. and Span. Colonies, 5,500,000; Dutch do. 1,600,000; Portuguese do. 2,000,000; Turkey and Smyrna, 5,000,000	
1787	23,250,268			1787 *Power Loom* invented. (") *Pennsylv'a Abolition Soc.* formed. (") Slavery excluded from N. W. Territory, including Ohio, Ind., Ills., etc. (1789) *Franklin* issues an appeal for aid to instruct free blacks.
1788	20,467,436			
1789	32,576,023		(1789) *Cotton crop* of United States, 1,000,000 lbs.	
1790	31,447,605			
1791	28,706,675	lbs. 189,316	*Imports* by Great Britain from Br. West Indies, lbs. 12,000,000; Brazil, - - - - - 20,000,000	
1792	34,907,497	138,328		1792 Emancipation by N. Hampshire.
1793	19,040,929	500,000		1793 *Cotton Gin* invented.
1794	24,358,567	1,601,760	*Cotton crop* of the United States 8,000,000 lbs.	
1795	26,401,340	6,276,300		
1796	32,126,357	6,100,000	*Cotton crop* of the United States 10,000,000 lbs.	
1797	23,354,371	3,800,000		
1798	31,880,641	9,330,000	*India*, the first imports from, 1,622,000 lbs.	

Years.	*Gr. Brit. Ann. Imp'ts & Conspt'n of Cotton, from earliest dates to* 1853, *in lbs.*	*United States Annual Exports Cotton to Gr. Britain and Europe generally.*	*Great Britain's sources of Cotton supplies other than the U. States, with total Cotton crop of United States.*	*Dates of Inventions promoting the growth and manufacture of Cotton, and of movements to elevate the African race.*
1799	43,379,278	lbs. 9,500,000	*Cotton crop* U. S., 20,000,000 lbs.	Emancipation by New York.
1800	56,010,732	17,789,803	*Exports* from—	(1804) Do. by New Jersey. *Cotton* consumed in U. States, 200,000 lbs.
1801	56,004,305	20,900,000	India, - - - lbs. 30,000,000 West Indies, - - - 17,000,000	*United States* exported to—
1802	60,345,600	27,500,000	Brazil, - - - - - 24,000,000 Elsewhere, - - - 7,000,000	France, - - - - - - lbs. 750,000 England, - - - - - - - 19,000,000
1803	53,812,284	41,900,000	. . . .	*Louisiana Territory* acquired, including
1804	61,867,329	38,900,000		the region between the Mississippi river (upper and lower) and the Mexican line.
1805	59,682,406	40,330,000	. . . .	U. S. exp't to France 4,500,000 lbs.
1806	58,176,283	37,500,000	*Cotton crop* U. S., 80,000,000 lbs.	
1807	74,925,306	66,200,000	. . . .	*Fulton* started his steamboat.
1808	43,605,982	12,000,000	. . . .	*Slave trade* prohibited by U. S, and Eng'd.
1809	92,812,282	53,200,000		(1808) *Cotton manufacture* established at Boston.
1810	132,488,935	93,900,000	. . . .	*Cotton* consumed in U. S., 4,000,000 lbs.
1811	91,576,535	62,200,000		
1812	63,025,936	29,000,000	*War* declared between United States and Great Britain.	Two-thirds of steam engines in Gr. Britain employed in cotton spinning, etc.
1813	50,966,000	19,400,000		(1813) U. S. exp't to France 10,250,000 lbs.
1814	73,728,000	17,800,000	. . . .	(1815) *Power Loom* first used in U. S.
1815	96,200,000	83,000,000	*Peace* proclaimed between United States and Great Britain.	(1816) *First st'mboat* crossed the Brit. Ch'l.
1816	97,310,000	81,800,000		" Power-Loom brought into general use in Great Britain.

1817	126,240,000	95,660,000	. . . .	*Colonization Society* organized.
1818	Total consump.[*] 109,902,000	92,500,000	*Cotton crop* U. S. 125,000,000 lbs.	
1819	109,518,000	88,000,000	. . . .	(1819) *Florida* annexed. (1820) Slave trade decl. piracy by Congress.
1820	120,265,000	127,800,000	*Exports* from— W. Indies, - lbs. 9,000,000	*Emigrants to Liberia* first sent.
1821	129,029,000	124,893,405	Brazil, - - - - 28,000,000	*Benjamin Lundy* published his "Genius of Universal Emancipation."
1822	145,493,000	144,675,095	India, - - - - 50,000,000 Turkey and Egypt, 5,500,000	
1823	154,146,000	173,723,270	Elsewhere, - - 6,000,000	U. States exp. to France 25,000,000 lbs.
1824	165,174,000	142,369,663	(1822) Cotton crop of United States, 210,000,000 lbs.	Do do. do. 40,500,000 lbs. (1825) New York and Erie Canal opened.
1825	166,831,000	176,449,907	. . . .	(1826) Creek Indians removed fr. Georgia.
1826	150,213,000	204,535,415		*Production* and *manufacture* of cotton now greatly above the *consumption*, and prices fell
1827	197,200,000	294,310,115		so as to produce general distress and stagnation, which continued with more or less in-
1828	217,860,000	210,590,463	*Cotton crop* U. S. 325,000,000 lbs.	tensity through 1828 and 1829. The fall of
1829	219,200,000	264,837,186	. . . .	prices was about 55 per cent.—*Encyc. Amer.* (1829) *Emancipation* in Mexico.
1830	247,600,000	298,459,102	. . . .	U. States exp. to France 75,000,000 lbs.
1831	262,700,000	276,979,784	. . . .	(1831) *Slave Insurrection* in Virginia.
1832	276,900,000	322,215,122	*Imports* by Gr. Britain from— Brazil, - - lbs. 20,109,560	*Wm. L. Garrison* declares war against the Colonization Society.
1833	287,000,000	324,698,604	Turkey and Egypt, 9,113,890	(1832) Ohio Canal completed.
1834	303,000,000	384,717,907	E. Ind. and Maur., 35,178,625 Br. W. Indies, - 1,708,764	Cotton consum. in France, 72,767,551 lbs. *Emancipation in W. Indies* commenced.
1835	326,407,692	387,358,992	Elsewhere, - - 964,933	(1834) James G. Birney deserted the Colonization Society.
1836	363,684,232	423,631,307	(1834) *Cotton crop* of United States, 460,000,000 lbs.	*Gerritt Smith* repudiates the Coloniz. Soc.

YEARS.	*Great Brit. Ann. Imports & Consumpt'n of Cotton, from earliest dates to 1853.*	*United States' Annual Exports Cotton to Gr. Britain and Europe generally.*	*Great Britain's sources of Cotton supplies other than the U. States, with total Cotton crop of United States.*	*Dates of Inventions promoting the growth and manufacture of Cotton, and of movements to elevate the African race.*
	. . . . .	. . . .	. . . .	(1835) U. S. exp. to Fr., 100,330,000 lbs.
	. . . . .	. . . .		
1837	367,564,752	444,211,537	*Imports* by Gr. Britain from— Brazil, - - lbs. 24,464,505	(1836) Cherokee and Choctaw Indians removed from Georgia, Mis'ippi, and Alabama.
1838	477,206,108	595,952,297	Turkey and Egypt, 5,412,478; E. Ind. and Maur. 40,230,064	(1837) American Anti-Slavery Soc. had an income of $36.000, and 70 agents commis'd.
1839	445,744,000	413,624,212	Br. W. Indies, - 928,425; Elsewhere, - - 5,377,215	(1838) Colonization Soc. had an income of only $10,900.
1840	517,254,400	743,941,061	*Cotton crop* U. S. 790,259,479 lbs.	Cotton consumed in U, S. 106,000,000 lbs.
1841	460,387,200	530,204,100	(1840) Imports by Great Britain from Brit. W. Indies, 427,529 lbs.	
1842	477,339,300	584,717,017	*Imports* by Gr. Britain fr. India, 1835 to 1839, annual average	
1843	555,214,400	792,297,106	57,600,000 lbs.	
1844	570,731,200	663,633,455	Do. do. 1840 to 1844, during Chinese war, 92,800,000lbs.	Value of cotton goods imported into the United States, $13,286,830.
1845	626,496,000	872,905,996	Do. from Egypt, 32,537,600 lbs.	*Texas* annexed.
1846	624,000,000	547,558,055		*Mexican War.*
1847	442,416,000	527,219,958	*Imports* by Gr. Britain from— W. I. and Dem. lbs. 3,155,600	*Gold* discovered in California. *New Mexico* and California annexed.
1848	602,160,000	814,274,431	Braz. and Port. Col. 40,080,400	U. States export to France 151,340,000 lbs.
1849	624,000,000	1,026,602,269	Mediterranean, - 11,604,000; East Indies, - - 91,004,800	Other continental countries, 128,800,000 lbs. *Cotton* consumed in U. S., 256,000,000 lbs.
1850	606,000,000	635,381,604	*Imports* by Gr. Britain from—	(1850) Value of Un. States cotton fabrics, $61,869,184.
1851	648,000,000	927,237,089	East Indies, lbs. 72,800,000; (1850) Do. do. 123,200,000	
1852	817,998,048	1,093,230,639	(1852) Do. do. 84,022,432	(1853) Value cottons import. $27,675,000.
1853	746,376,848	1,111,570,370	(1853) Do. do. 180,431,496; *Cot. crop* U. S. 1,600,000,000 lbs.	U. S. exp. to England, - 768,596,498 lbs. Do. do. to Continent, - 335,271,064 lbs.

NOTE.—Our commercial year ends June 30: that of England, January 1. This will explain any seeming discrepancy in the imports by her from us, and our exports to her.

N. B. In 1781 Great Britain commenced re-exporting a portion of her imports of Cotton to the continent; but the amount did not reach a million of pounds, except in one year, until 1810, when it rose to over eight millions. The next year, however, it fell to a million and a quarter, and only rose, from near that amount, to six millions in 1814 and 1815. From 1818 her *consumption*, only, of cotton is given, as best representing her relations to slave labor for that commodity. After this date her exports of cotton gradually enlarged, until, in 1853, they reached over one hundred and forty-seven millions of pounds. Of this, over eighty-two millions were derived from the United States, and over fifty-nine millions from India.

TABLE II.

Tabular Statement of Agricultural Products, Domestic Animals, etc., EXPORTED *from the United States; the Total Value of Products and Animals* RAISED IN THE COUNTRY; *and the Value of the portion thereof left for* HOME CONSUMPTION AND USE, *for the year* 1853. *See Patent Office Report; Abstract of Census; Rep. Com. Nav., etc.*

	Value of Exports.	Total Value of Products and Animals.		Value of portion left for home consumption.
Cattle, and their products,........	$3,076,897	Cattle,	$400,000,000	$396,923,103
Horses and Mules..................	246,731		300,000,000	299,753,269
Sheep and Wool....................	44,375	Sheep,	46,000,000	45,955,625
Hogs, and their products.........	6,202,324	Hogs,	160,000,000	153,797,676
Indian Corn and Meal............	2,084,051	Corn,	240,000,000	237,915,949
Wheat Flour and Biscuit.........	19,591,817	Wheat,	100,000,000	80,408,183
Rye Meal...........................	34,186	Rye,	12,600,000	12,565,814

Other Grains, and Peas and Beans	$165,824	54,144,874	53,979,050
Potatoes	152,569	42,400,000	42,247,431
Apples	107,283	(1850) 7,723,326	7,616,043
Hay, averaged at $10 per ton	....	(1850) 138,385,790	138,385,790
Hemp	18,195	4,272,500	4,254,305
Sugar—cane and maple, etc	427,216	(1850) 36,900,000	36,472,784
Rice	1,657,658	8,750,000	7,092,342
Totals	$33,809,126	$1,551,176,490	$1,517,367,364
Cotton	$109,456,404	$128,000,000	$18,543,596
Tobacco, and its products	11,319,319	19,900,000	8,580,681
Totals	$120,775,723	$147,900,000	$27,124,277

TABLE III.

Total Imports of the more prominent articles of Groceries, for the year ending June 30, 1853; *specifying, also, the re-exports, and the proportions from Slave-Labor Countries. See Report on Commerce and Navigation.*

Coffee, Imported	Value, $15,525,954	lbs. 199,049,823
" Re-Exported	1,163,875	" 13,349,319
" Slave-Labor production	12,059,476	" 156,108,569
Sugar, Imported	$15,093,003	" 464,427,281
" Re-Exported	819,439	" 18,981,601
" Slave-Labor production	14,810,091	" 459,743,322
Molasses, Imported	$3,684,888	gals. 31,886,100
" Re-Exported	97,880	" 488,666
" Slave-Labor production	3,607,160	" 31,325,735
Tobacco, etc., Imported	$4,175,238	....
" Re-Exported	312,733	....
" Slave-Labor production	3,674,402	....

ERRATA.

Page 36, 12th line from top, for "grapes of gall," read, "their grapes *are grapes* of gall."

Page 38, 12th line from top, for "*explaining*," read, "*exclaiming*."

Page 68, 8th line from top, for "*for more*," read, "for *no* more."

www.ingramcontent.com/pod-product-compliance
Lightning Source LLC
LaVergne TN
LVHW010742120826
845150LV00009B/1398
* 9 7 8 1 4 2 5 5 1 7 4 7 2 *